A Walk through Fuerteventura

Qué nombre tan sonoro, alto y significativo!
Fuerteventura? Es decir, ventura fuerte.
What a name, so resonant, high and significant!
Fuerteventura? That is to say, great venture.

Miguel de Unamuno, 1924

No man should go through life without once experiencing healthy, even bored solitude in the wilderness, finding himself depending solely on himself and thereby learning his true and hidden strength.

Jack Kerouac

No hay camino, se hace camino al andar.
There is no road, the road is made by walking.

Antonio Machado

Caminar es atesorar
To walk is to gather treasure

Old Spanish saying

A Walk through
Fuerteventura

David Collins

HOUBARA HOUSE

British Library Cataloguing in Publication Data
A catalogue record for this book is available from the British Library

ISBN 978-0-9571505-0-8

Typeset by Amolibros, Milverton, Somerset
This book production has been managed by Amolibros
www.amolibros.com
Printed and bound by T J International Ltd, Padstow, Cornwall, UK

About the author

David Collins is a lifelong naturalist and a professional ecologist. He graduated from Royal Holloway College, University of London, where he completed an honours degree in Biological Sciences in 1979. This was followed by two years studying the behaviour and ecology of the rare and endangered Houbara Bustard in Fuerteventura for which he was awarded a Master of Philosophy degree. He has subsequently visited the island on many occasions to lead wildlife holidays, to study bird migration and wildlife, and on holidays with his family. He is co-author of the definitive guide to birdwatching in the Canary Islands and has also written popular articles on the Canary Islands for birdwatchers.

His varied career began with the Royal Society for the Protection of Birds, where he contributed both to the society's research and conservation planning activities. Since then he has worked as an environmental consultant, as senior ecologist for major infrastructure projects, and as environmental advisor both to the Environment Agency and the Department for Environment Food and Rural Affairs.

He is a Chartered Environmentalist and a founder member of the Chartered Institute of Ecology and Environmental Management (CIEEM), of which he is a past member of Council.

Contents

List of Boxes

List of plates

Book Cover – The volcanic cone of Montaña Arena from the south.

Colour photographs, facing page 18

Black & white photographs, facing page 178

Acknowledgements

I am very grateful to my friends Kate Watters and Chris Durdin who were kind enough to read through the manuscript and provide constructive editorial comments. I have no doubt that the book is significantly better for their efforts. I am also grateful to my wife Eiluned, who provided helpful comments on various drafts, but more importantly provided the moral and practical support needed both to undertake the walk and complete the book.

I am grateful to Ray Purser for permission to use his superb photograph of the Cream-coloured Courser. All other photographs are by the author. Thanks also to Richard Unthank who digitised the maps.

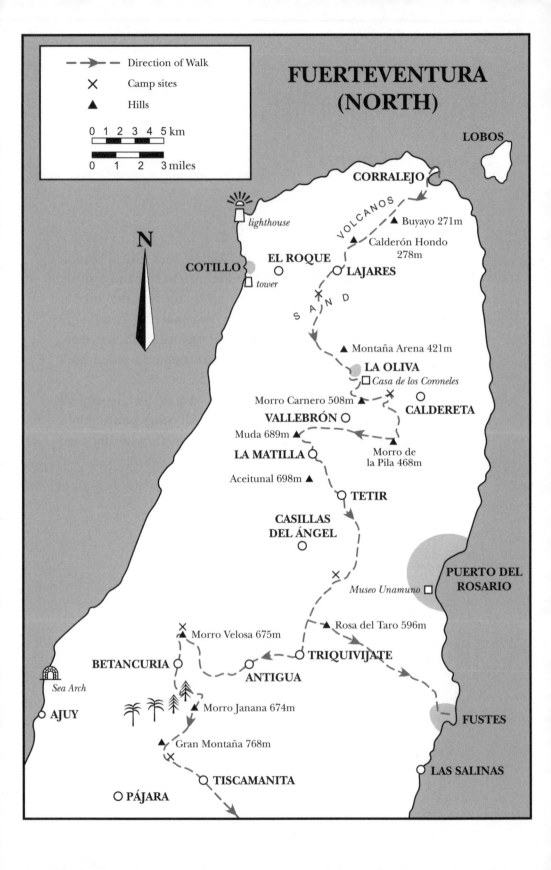

FUERTEVENTURA (NORTH)

LOBOS

CORRALEJO

VOLCANOS

Buyayo 271m

Calderón Hondo 278m

lighthouse

COTILLO

EL ROQUE

LAJARES

tower

S A N D

N

Montaña Arena 421m

LA OLIVA

Casa de los Coroneles

Morro Carnero 508m

CALDERETA

VALLEBRÓN

Muda 689m

Morro de la Pila 468m

LA MATILLA

Aceitunal 698m

TETIR

CASILLAS DEL ÁNGEL

PUERTO DEL ROSARIO

Museo Unamuno

Rosa del Taro 596m

Morro Velosa 675m

BETANCURIA

TRIQUIVIJATE

ANTIGUA

Sea Arch

AJUY

Morro Janana 674m

FUSTES

Gran Montaña 768m

LAS SALINAS

TISCAMANITA

PÁJARA

Legend

⟶ — — Direction of Walk
✕ Camp sites
▲ Hills

0 1 2 3 4 5 km
0 1 2 3 miles

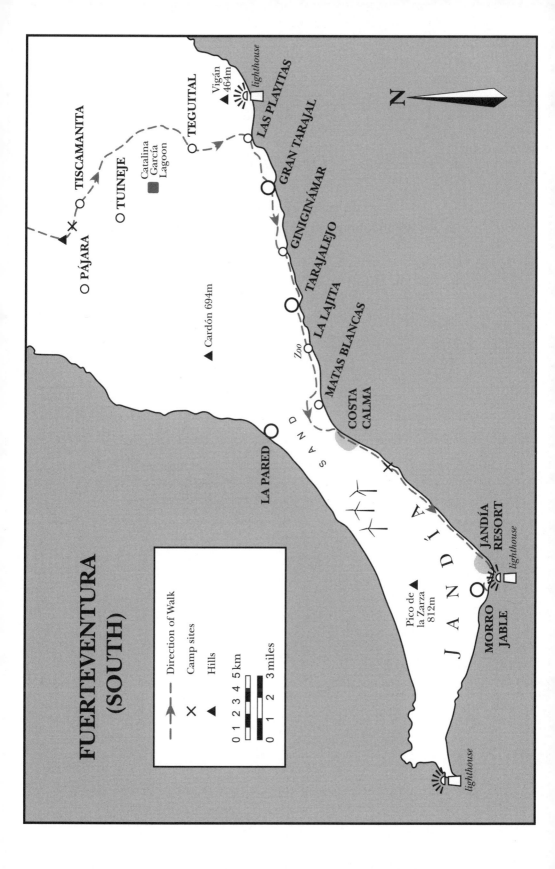

FUERTEVENTURA (SOUTH)

N

Direction of Walk
X **Camp sites**
▲ **Hills**

0 1 2 3 4 5 km
0 1 2 3 miles

TISCAMANITA

TUINEJE

PÁJARA

TEGUITAL

Catalina
García
Lagoon

Vigán
464m

lighthouse

LAS PLAYITAS

GRAN TARAJAL

GINIGINÁMAR

TARAJALEJO

▲ Cardón 694m

LA LAJITA

Zoo

MATAS BLANCAS

COSTA
CALMA

LA PARED

S
A
N
D

J
A
N
D
Í
A

Pico de
la Zarza
812m ▲

JANDÍA
RESORT

lighthouse

MORRO
JABLE

lighthouse

Dedication

In memory of my father,
who always encouraged and supported my interest in natural
history.

ONE

Introduction

Fuerteventura is the second largest of the Canary Islands. This book describes my attempt to walk from one end of it to the other, incorporating some of its highest mountains. As far as I know, nobody had tried to do this before. It is also an account of the island's natural history and places of historic interest, based on knowledge built up over more than three decades of visits. My aim has been to write a readable account of the island, and to provide more detail than can be found in guide books. With the exception of one or two long out-of-print books, there is nothing similar in the English language.

I first set foot in Fuerteventura on 30th March 1979. Little did I know then what an impact the island would have on the rest of my life. I was an impressionable twenty-one-year old student, taking part in an expedition organised by the International Council for Bird Preservation (ICBP), now BirdLife International. Our task was to estimate the population of the Houbara Bustard, a rare bird of the desert fringes.

It seems incredible, embarrassing almost, that we called ourselves an expedition, as today Fuerteventura is a popular tourist destination. But in those days Fuerteventura was a harsh land of poverty and open desert landscapes. Life on the island was harsh and facilities were basic except in the main towns and the few isolated tourist hotels. Villages had no mains electricity, and goats were still the mainstay of a predominantly agrarian economy. Three weeks on a sparsely populated semi-desert island studying rare birds was certainly wild adventure for me.

Now, life on the island is generally a great deal more comfortable. Unfortunately, though, development has inevitably come at a cost. All too often, what used to be an awe-inspiring wilderness of naked lava is now covered in apartments. The drive to accommodate as many tourists as possible has sometimes overruled sensitivity to the natural environment.

There has also been a dramatic change in the availability of water. Where once there was desperate shortage and every drop was used carefully, tourism has required the development of expensive desalination facilities, which have made it possible to lavish water on sub-tropical landscaping on a scale that would previously have been unthinkable. Some of the islands resorts now even have golf courses.

Despite the head-long rush to develop the island for tourism, however, wilderness and tranquillity can still be found, and the island's fascinating natural history is still largely intact. I hope that those who read this book will share the pleasure that the island and its wildlife have given me over more than three decades, and will perhaps be encouraged to visit the island to explore it themselves.

★

Fuerteventura is the closest of the Canary Islands to Africa. At its nearest point it is about 100 kilometres from the western coastal fringe of the Sahara. Like all the islands it is volcanic in origin, but it is older than the others, and much older than the western islands. The violent volcanic activity that created the island began twenty million years ago, and lasted for eight million years. By then, the island was something like modern day Tenerife, with mountains 3,000 metres high. But today the island is just the eroded remains of those mountains. Its highest point, at Pico de la Zarza, is only 812 metres above sea level (807 metres according to older maps).

After a long period of relative quiet, there was a resurgence of volcanic activity five million years ago, with smaller eruptions continuing over the last three million years. This recent period of volcanism has, however, been on a much smaller scale. The volcanic cones and lava fields, mainly in the north and centre of the island, bear testimony to that recent period of upheaval and form a distinctive part of the landscape. In places, solidified lava flows are evident, but more often volcanic bombs cover the lava in jagged rocks. The cones themselves are formed by accumulation of pyroclasts: volcanic ash that is mostly deposited down-wind, so one side of each cone has a higher rim than the other. These lands of lava, rock and ash are known as *malpais*, literally 'badlands': a good name for territory that is treacherous to cross and cannot be farmed.

There is an interesting museum about the volcanic history of the island at Cueva del Llano, between Corralejo and Villaverde (Box 1).

★

Above all else, Fuerteventura is a thirsty island. Most areas receive only 50 – 150 mm of rain per year on average, and the coastal fringe is even dryer. Almost all of the rain falls between November and April, and there is virtually none at all in summer. There is also much variation between years. In the wettest years there may be 500 mm in the high mountains and even very arid areas such as the east coast may get more than 200 mm. In contrast, even the highlands receive less than 50 mm in the worst years. In the past, years of low rainfall led to almost complete crop failure, and both livestock and people suffered terribly.

In keeping with its aridity, Fuerteventura has a desert landscape. Its hills and plains are generally bare; the only obvious vegetation over the vast majority of the island is a sparse scattering of dull green saltwort (Box 2). In the few areas where there is enough moisture to allow cultivation, palms and fig trees form oases in the harsh landscape. After the intermittent heavy rains the hillsides flush green with annual grasses, and for a few weeks the landscape is dotted with the colourful flowers of annual herbs.

★

Today, travel to Fuerteventura is simple and easy. There are direct flights from all over Europe, and passenger ferries operate a regular service across the shallow waters between Lanzarote and Corralejo, the main resort at the northern end of the island. A hundred years ago, getting to Fuerteventura was much more difficult. The first impressions of those few seasoned travellers who made it to the island must have been very different. In the last year of the nineteenth century, when the pioneer bird photographer Henry Harris visited Fuerteventura, recently introduced steam ships provided the inter-island service. His comments about travel to the island in years before the steamships came into service are telling.

> "The journey now is very different to what it was some ten years ago, when the enthusiastic naturalist – for no one not interested in natural history would go to Fuerteventura – must take his chance in the rough seas which are often to be encountered between these islands, trusting himself and his belongings to some antiquated schooner which might very well take a week over the voyage."

Fourteen years later, in May 1913, during his British-Museum-sponsored expedition to the Eastern Canary Islands, David Bannerman, who was one of the leading ornithologists of the day, joined a similar small steamer from Las Palmas in Gran Canaria. When his ship arrived at the island's capital (then Puerto Cabras or 'Goat Port', the pre-tourist era name for Puerto del Rosario) it anchored offshore and they were taken ashore in small boats. Both baggage and passengers were then loaded onto camels. For the latter purpose wooden frames were set on the camels so that one passenger could ride on either side.

When the island began to attract tourists in the 1960s, the airport south of Puerto del Rosario had yet to be built. Planes from Las Palmas and Arrecife landed at the old military airport at Los Estancos, half-way between Puerto del Rosario and Tetir. In those days many travellers still used the inter-island ferry, which took around two and a half hours to ply between Arrecife and Puerto (as the capital is generally called). This was cheaper, but not a good option for those without good sea-legs, as the sea to the east of the islands is deep and there is often a big swell.

★

In common with almost everyone who visits the island these days, my first view of Fuerteventura was from the air. In the late 1970s, the airport at Puerto was really just an airfield with a short runway that could only accommodate small planes. There were no direct flights from England then, so we had flown to Arrecife on the nearby island of Lanzarote, and had to stay there overnight before catching the plane to Fuerteventura. True to the expeditionary tradition (and to save money), we camped on volcanic grit by the roadside somewhere inland from Arrecife. The following day we made the short hop to Fuerteventura in a small and rather ancient Fokker. From above, the island looked utterly barren, just rounded mountains with a few volcanic cones, and apparently no signs of life save for a few low buildings and an empty road. I had never been anywhere remotely like it before.

We touched down at the smallest airport I had ever seen. Beside the runway was a large shed that served as the airport building, complete with a dingy bar where a few locals whiled away the hours in a smoke-filled fug. The utter barrenness of the plains near the airport certainly didn't inspire me, and this is certainly the dreariest part of the island. Neither, unfortunately, was my first day on Fuerteventura a particularly auspicious one. After landing, we took taxis into Puerto del Rosario, where we met up with three local bird experts who were part of the Grupo Ornithologico Canario. They had arrived earlier from Tenerife on the inter-island ferry. To celebrate our arrival on the island and to get to know each other a little better we had a drink in a bar near the quay. In the excitement of it all, I was stupid enough to leave my day sack at the bar, and that was the last I saw of it. Fortunately, I lost nothing of particular importance: my passport and the little money I had was in my jacket, and my trusty binoculars were, as ever, round my neck.

On that first day in Fuerteventura I was immensely excited at the prospect of spending three weeks in a place that was so utterly different to anything that I had known before, watching Houbara Bustards and a range of other desert birds that I had previously only dreamt of seeing. I had no idea, of course, that twenty-nine years later I would be back there with the intention of walking from one end of the island to the other.

★

Box 1: *Cueva del Llano*

Anyone who is interested in the volcanic history of the island would do well to visit the museum at Cueva del Llano. This is to the west of the road between Corralejo and La Oliva. The museum is at the entrance to a hollow underground lava tube, more than half a kilometre long. Regular tours take groups of visitors along part of the tube. Most tours are in Spanish, but there are English tours at certain times, and the museum provides information in English.

The cave is home to a unique spider-like creature *Maiorerus randoi*, which is blind and un-pigmented as a result of living for hundreds of thousands of years in the dark. It is so unlike anything else in the world that it has been given its own genus, dedicating it to the Majoreros, as the people of Fuerteventura are called. But don't worry if you are an arachnophobe, you won't see it down there: it is rare and confined to the humid depths of the lava tube where visitors are not allowed to go.

Box 2: Desert shrubs

Most of the desert shrubs that are scattered across the dry hillsides and plains are species of saltwort. The commoner species include *Salsola vermiculata*,[1] with its beautiful red-winged fruits, and the dull grey-green *Chenoleoides tomentosa*. There are two other common shrubs scattered amongst the saltwort. The first is a member of the daisy family, and is nothing more than a tangled mass of harmless dull green spines that are too weak to damage the skin. This is *Launaea arborescens*, locally known as *aulaga*. It has a characteristic strong musty smell, is virtually leafless and is one of the most characteristic plants of the island. After rain it produces a mass of yellow flowers not unlike a sowthistle. The other is *Lycium intricatum*, a member of the nightshade family locally known as *espino* (literally 'the spiny one'). It is a sprawling plant with nasty spines rather like those of blackthorn. The plant is leafless during dry periods, but produces small leaves and purple flowers after rain. These develop into little red berries that are quite pleasant to eat, and are a valuable source of food for some of the island's birds. Another plant that is very obvious beside roads and around cultivated areas is the Shrub Tobacco *Nicotiana glauca*, an introduced plant with a woody stem several metres long on which are borne clusters of tubular yellow flowers.

---

[1] Plants mentioned in the text are listed in Appendix 2, which provides the common names (where available), and brief notes on appearance and distribution in Fuerteventura

Two

The Fuerteventura Houbara Expedition

My inclusion in the Fuerteventura Houbara Expedition certainly had an element of luck to it. At the time I was a third-year ecology student at Royal Holloway College, London. Our plant geography course included a week studying the flora of Tenerife, which gave me a basic grounding in the vegetation of this fascinating island in the Canary archipelago. The bustard enthusiasts within the ICPB, who were trying to get a group of experienced birders together for the Fuerteventura Houbara expedition, needed someone who could also provide botanical support. A keen birder I most definitely was, and although the plants of Tenerife and Fuerteventura are very different, my knowledge of Canary Island plants was deemed to be sufficient for me to be accepted into the team. I was thrilled, partly at the thought of seeing exciting new birds, and partly because of my love of wild places.

Our base for the three weeks was at La Oliva, a large village of low, whitewashed houses that serves as the administrative capital of the north. It is halfway between the east and west coasts, on the southern edge of a great lava field of jumbled rocks. Rising above the village is Montaña Arena, which is one of the island's more impressive volcanic cones.

We were lodged in empty buildings with no electricity on the very edge of Arena's lava field. We slept in sleeping bags on the floor of the largest room, and at night, after an exhausting day out in the sun and wind, wrote up our notes by the light of kerosene lamps.

The buildings had previously been the centre of the government-run sisal industry. Sisal, which is used to make rope, is the dried fibre of the agave, an American plant. At the time of our expedition, agave plantations still covered great swathes of land around La Oliva. The lorry that been used to collect the agaves and take the fibre to the port for shipment was rusting away in the courtyard. Clearly the commercial agave experiment had not been a lasting success (Box 3).

During our three weeks in Fuerteventura, we spent every day walking the plains in search of Houbaras. The Houbara is quite a big bird, about the size of a small goose but with rather long legs, slender build and an elegant posture (Plate 1). It is amazingly good at avoiding detection, and it required a lot of concentration to spot one. We would select a different plain each day, and head out early in the morning with a packed lunch, returning hot and tired in the late afternoon, having walked all day under a burning sun. After about a week the backs of my hands became so sunburnt that they blistered, and I spent the rest of the expedition with both hands wrapped in bandages. This earned me the nickname 'Rocky' after my resemblance to the boxer Rocky Marciano, or at least his hands.

Much easier to spot than the Houbara, and quite trusting, is the Cream-coloured Courser (Plate 3). If the Houbara is the king of the plains, the dainty courser is surely the princess. These plover-like birds are about the size of a thrush, but sleek and graceful with a slightly down-curved bill. They are not cream but pale sandy-brown with striking pastel blue, white and black lines on the head. When they stand still their general colouration is so similar to the soil that they are almost invisible. But unlike the bustard they never stand still for long. After a few moments they lean forward to dash a few metres before stopping to look around again, standing with a high posture that gives them the best chance of spotting a beetle or ant. The juvenile is a little paler than the adults, with a subdued head pattern and speckling for even better camouflage.

Although the island does not have a wide range of birds, there are a few that are common everywhere, including Spanish Sparrow, Collared Dove (a relatively new arrival), Lesser Short-toed Lark and Berthelot's Pipit (Box 4). The pipit is unique to the Canary Islands and Madeira. The status of all birds that have been reported in Fuerteventura is provided in Appendix 1.

★

The long days out on the plains were certainly hard work but I loved every minute. We saw the island in a way that no tourist would. We often passed goatherds contemplating their flocks and with no choice but to be at one with their surroundings. Sometimes we came upon wonderful sandy coves with no one in sight, where we could eat lunch with the heavy surf pounding the shore, and splash in the waves to cool down. More often we lunched behind walls of lichen-crusted lava that smelt of goats. These walls gave shelter from the desiccating wind but not the scorching sun.

In three weeks of walking the plains, I personally saw only a handful of Houbaras. Indeed the whole expedition only found forty-two. We estimated the total population of the island to be no more than 100, although later surveys have confirmed that the population is considerably larger.

★

At the end of the expedition I returned to the more important, if less exotic task of completing my degree. And that could have been that, as far as my interest in Fuerteventura was concerned. However, it transpired that the ICBP was thinking of funding someone to study the Houbara in more detail. I couldn't think of anything I would rather do than spend the next few years studying a rare bird somewhere abroad, so I made my interest known. To my surprise, nobody else put their name forward, so I was offered the post. Although it was a daunting prospect, it was simply too good an opportunity to miss, and I therefore returned to the island to study the Houbara in December 1979. I stayed until April 1980, returning for a second period of study between January and April the following year.

★

Box 3: Agaves in Fuerteventura

The agave plantations around La Oliva, as elsewhere on the island, were planted in the 1930s and '40s. The plantations are of two species, *Agave fourcryoides* and *Agave sisalana*. The fibre is made from the tough, leathery leaves, but the most striking feature of the agave is its flower. After some years of growth it produces a flower spike 8m high over a period of a few weeks: then it dies.

The third species of agave that is common on the island is *Agave americana*. It is planted to mark the limits of estates and gardens, forming eccentric but entirely impenetrable 'hedges'. It is a strikingly sculptural plant with a rosette of gracefully down-curved, blue green leaves about a metre long. The leaf margins are barbed with stiff spines, and at the end of each there is a nail-like spike that can cause serious injury if brushed against, as I once found out.

Box 4: Common small birds

The three most ubiquitous small birds on the island are Berthelot's Pipit, Lesser Short-toed Lark and Spanish Sparrow.

Berthelot's Pipit (Plate 37) is found only in the Canaries and Madeira, and is very common in Fuerteventura. It is the small streaky brown bird that you see running around on the plains, on the edge of villages and in tourist complexes. Its song is undistinguished, just a repetition of its simple, chirpy 'tsiree' call, often given in song flight. If disturbed from its nest it will often flutter along the ground as if it has a broken wing in order to distract you from its eggs.

In contrast to the pipit, the Lesser Short-toed Lark, like its larger relative the Skylark, is a real songster. Where the habitat is good, its song fills the air. It is a good mimic, incorporating snatches of other birds' songs in its own, and seems to particularly like imitating the Linnet, which is fairly common, and the much scarcer Corn Bunting. During the dry season the larks roam the plains in busy flocks that can be hundreds strong.

The Spanish Sparrow in the only sparrow in Fuerteventura. The male is a handsome bird with a bold black waistcoat. Linnets are also fairly common.

THREE

Houbara Bustards

I hadn't really thought through the consequences of going out to Fuerteventura alone to study the Houbara, but that is what I found myself doing towards the end of 1979. This time I flew to Las Palmas in Gran Canaria, and then on to Fuerteventura, arriving there on 2nd December. Initially I stayed at the same building in La Oliva, which was now equipped as a Biological Station. Fernando Domingues, a native of Tenerife, who had been part of our expedition, was in residence there with his family whilst he studied the introduced Barbary Ground Squirrel. At first I did not have permission to stay in the building, so had to make do with sleeping on the concrete floor in one of the outhouses. I had a sleeping mat so this was not too uncomfortable. The problem was the mice! Just as I was dropping off to sleep, they would come out and I would hear them scampering around. If I turned on the torch I would often surprise them inches from me. But they never nibbled at either me or the sleeping bag, and I soon learnt to ignore them. Having done so, I slept soundly enough there until, after a couple of weeks, permission was given for me to move indoors.

A far bigger problem than the mice were the Houbara Bustards, or rather my inability to find them, or indeed to keep them under observation for any period of time once I did find them. In the first two weeks I saw only one, and that was nearly ten kilometres away and I didn't have any transport. How exactly was I supposed to study a bird I couldn't find? Giving up and going home was one option that occurred to me at this point. Instead I invested some of my very little money (the enterprise was privately funded) in a small-wheeled bicycle. It had no gears and was almost impossible to power up the steep hills, but it did give me access to the area west of the village of Tindaya, where I had seen that one bird in the first week. Here I could watch a group of three birds that came to feed in some cultivations every morning and evening.

Fernando and his family went home at Christmas and I was left alone for an intensely lonely week. On top of the loneliness I had one other problem: lack of food. I had marked out a particular day to go into Puerto to shop for Christmas, only to find that the bus had already stopped running for the week. The shops in La Oliva stocked only the most basic foods, so this meant a subsistence diet of fried luncheon meat, rice and tomatoes through the festive season, except for a box of butterscotch that my parents had sent me as a Christmas present. Still, there was the novelty of being somewhere warm at Christmas, so I determined to make the most of it. On Christmas Day I sat outside and sucked butterscotch in the warm sunshine in a somewhat contemplative mood.

My stay at the Biological Station lasted only until mid-January 1980. Then with the assistance of one of the local hunting guards, who drove me to various possible apartments and acted as my interpreter, I found somewhere to rent in the centre of the village. This was to be my home during the rest of my bustard studies. It wasn't much: a single storey building with a flat roof, like almost every building in the village at the time. It had one large room (containing two beds), a kitchen at one end and a separate toilet and shower room. The toilet blocked periodically and my landlord had to come down and sort it out by taking the floor up. The water supply was do-it-yourself: a water tank on the roof had to be filled from a covered underground tank just outside the door. A ladder gave access to the roof and to fill the water tank required a very large number of bucketfuls to be taken up there. Over the kitchen there was an open-sided vent through which large quantities of red Saharan dust came in when the wind blew from the east. At such times a thick layer of dust had to be swept from the kitchen table before it could be used at meal times. The opening also allowed easy access for the mosquitoes that bred quite successfully in the water tank. The house did at least have electricity, although La Oliva was on a generator in the first year, and power was only available for a short time in the morning and evening.

My new accommodation was rented from a well built, swarthy man who lived at the far northern end of La Oliva. His name was Pepe Suarez, though he liked me to call him Don Pepe, the prefix suggesting a man of importance. It was some time before I realised that he was illiterate, as were many of his generation. He was a friendly man, and whenever he came round for the rent or to undertake some maintenance task, we always tried to exchange a few words.

One Sunday morning when I had a visitor staying with me, we had reason to visit Don Pepe at his house. He and a friend were having a 'feast', cooking something in a filthy looking frying pan in the yard. We were invited to join them. They explained that the unattractive objects in the frying pan were cubes of goat fat, and the pale substance in a large bowl was *gofio*. This is a traditional island dish, made from whatever grain or pulse is available, and kneaded into small balls. In his excellent book on the island, John Mercer explains that *gofio* was the staple diet of the pre-Conquest islanders. Today it is made from wheat, maize, barley or pulses, and is often a mix of several different ingredients. In the past, wild plant seeds were used when no other food was available. The grain is roasted and then ground to make flour. It seemed to me to have no flavour at all, and after a few balls of the stuff and some cubes of goat fat we made our excuses and left.

The house that I rented was opposite the church and the square, which in those days was a quiet space that was little used. However, just after midnight on 2nd February 1980, I was woken by several terrifically loud bangs. I rolled over and tried to go back to sleep, but when the banging began again, curiosity got the better of me. I poked my head out to see what was going on. To my utter surprise the street was full of people, and the sky was lit with fireworks. The Fiesta of Our Lady of Candelaria had started. Over the following days the square was full of stalls selling sweets and trinkets, and there were sometimes musicians dressed in traditional costumes and dancing in the square. The parade itself was a very sombre affair with a strong military attendance. Looking back, it saddens me to realise that I was so isolated from village life that I had no idea when its most important event of the year was about to happen.

★

Having settled into my new home, life became somewhat monastic. I spent every day except Sunday roaming the desolate plains in the sun and wind, trying to observe the Houbaras. Occasionally I passed a goatherd and we would exchange a few words, although my Spanish was rudimentary. I would ask whether they had seen Houbaras and what they knew of them, and maybe pass comment on the weather. I understood some of what they said, but must have missed much. In the evenings I wrote up my notes and studied a school Spanish text book. There was no television or radio, though I had a few music tapes. I could buy bread and other essential items in the village but the highlight was my weekly trip on the bus into Puerto for provisions and a cup of coffee in one of the bars. I was certainly lonely, but the solitude also brought a calmness I had not known during teenage and university years.

Fortunately, people came out to visit me from time to time. Without them, I'm not sure I would have been able to stay the course. Amongst the early visitors was my girlfriend Rosie. One day we went for a walk over the plains to the west of La Oliva, I think to give her an idea of what my work was like. To my delight and utter disbelief we chanced upon a displaying male Houbara. It turned out that there was a good population of the birds much nearer home than I had realised. From then on I abandoned the Tindaya birds and spent the rest of my time studying the population of birds that we had stumbled on. At last I was in business.

The equipment I had to study the bustards was as meagre as my living was spartan. My tools were a pair of Zeiss binoculars, a lightweight tripod that I could set the binoculars on to steady them during periods of prolonged observation, a collapsible canvas stool, a hand-held tape recorder for recording my observations, and my trusty bicycle. The equipment all went into a day sack, although it was a bit of a squeeze once lunch, water and everything I needed for the day was in it, and the tripod legs habitually stuck out the top a little.

One day, I took some visiting friends to Betancuria in the mountainous centre of the island. The owner of the restaurant we dined in recognised me. This surprised me as Betancuria and La Oliva are some way apart, and it hadn't occurred to me that there was much communication between the two. According to our host, the rumour amongst the good people of La Oliva was that I was a spy who sent information back to England, the tripod legs being the antennae of a radio. It seems they also thought my day sack bulged because it was stuffed full of money! If only. Of course I was living in a tight-knit community where strangers would automatically be treated with some suspicion, especially if they kept themselves to themselves, as I certainly did.

I quickly learnt that cars provide excellent hides for watching Houbaras. Whilst it is necessary to be at least 500 metres from a Houbara to observe normal behaviour when on foot, it is possible to get much closer in a car. For this reason, my visitors and I drove ordinary cars over impossible rocky terrain and sandy tracks that really required 4-wheel drive vehicles. As a consequence, we got an awful lot of punctures.

Normally, though, I did not have a vehicle, and was forced to watch Houbaras at long range. My observations of displaying birds were often made from almost a kilometre away, at which distance they were little more than dots. This required great concentration and persistence. Another early visitor was Paul Goriup, who had led the Fuerteventura Houbara Expedition and was now advising me on my studies. He came up with the innovative idea of creating a bird hide from a large cardboard box. We went into Puerto and after a while tracked down such a box. I enjoyed the creative effort of cutting a hinged door and observation flaps, together with strings for tying these open or closed. This having been achieved, I took the box out to the study area and set it up near where one of the males was in the habit of displaying. I spent a good few hours in this make-shift hide over the next few days, but I never did see any Houbaras from it. Then one morning I arrived there to find a camel sitting on it. The animal had evidently eaten its fill of the box and then decided the remnants made good bedding.

The following year I tried a proper canvas bird hide, lent to me by the ICBP. Once it was in place I pitched my tent next to it so I could get into it at dawn without the Houbaras seeing me. This didn't work either. I decided to leave it in place for a few days before trying again. No problem with camels this time, but when I returned I was dismayed to see that the Spanish Foreign Legion had set up camp around it. I didn't fancy marching into their camp to retrieve it. Unfortunately, when they left it had gone, and I suspect they thought it was a latrine and had been using it as such. At that point I gave up on the idea of using hides.

★

Despite the lack of resources and the fiendishly uncooperative behaviour of the birds themselves, I did manage to learn quite a lot about the Houbaras. It turned out that they did not form monogamous pair bonds, as described in all the standard textbooks. Instead, I found that females had sole responsibility for incubating the eggs and rearing the young.

The males have perhaps the most outrageous display of any bird outside the tropics. They display from prominent locations, a slight rise perhaps where they are as visible as possible over the surrounding plains. Each male has a favourite spot, an arena, where he spends the greater part of the early morning and evening displaying. The males I watched had arenas at least 500 metres apart. At first a male who is intending to display stands motionlessly, posed stiffly erect. Then he begins to erect the black and white feathers on his head and neck. Suddenly, his neck kicks back and his breast feathers envelope the front half of his body, so that from a distance he seems to have turned into a white ball. At the same instant he begins to run this way and

1 Left: Houbara Bustard.
2 Above: Houbara Bustard nest.

3 Above: Cream-coloured Courser.
4 Right: Male Fuerteventura Chat.

5 *Ruddy Shelduck.*

6 *Above: The ridge leading to Muda.*

7 *Left: The amazing carvings on the front of Pájara church.*

8 *Above: View to the south from Rosa del Taro with* Kleinia neriifolia *in the foreground.*

9 *Left:* Limonium papillatum *near Toston lighthouse.*

10 *Below: The 'new' harbour in Cotillo with eighteenth-century defence tower. The former island of Roca de la Mar is on the right.*

11 Above: Pulicaria
canariensis *on Roca
de la Mar, Cotillo.*

12 Right: Lotus
lancerottensis.

13 Below: Kickxia
heterophylla – *a
common plant
throughout the island.*

14 *Left: Plain Tiger butterfly.*

15 *Below:* Mesembryanthemum crystallinum – *with showier flowers than the more common* M. nodiflorum.

16 *Bottom: View to the north from the 'camp site' at Morro Velosa. The shrub in the foreground is the endemic* Asteriscus sericeus.

17 *Descent to Betancuria, with* Euphorbia regis-jubae *and Agave.*

18 *The pilgrims' path curves round an Agave clump on the descent from Gran Montaña, with the village of Tiscamanita in the distance.*

19 The lagoon at Catalina Garcia.

20 Euphorbia canariensis *at Cardón.*

21 *Above: La Lajita church on Easter Sunday.*

22 *Left: The sheer cliffs of Jandía are home to rare plants.*

23 *Below:* Caralluma burchardii.

that, often around a particular group of stones or shrubs. Then, after a few seconds of this madness, he stops as abruptly as he had started, and his feathers begin to relax. He stops long enough to check for danger, and no doubt to get his breath back, before repeating the display. He carries on like this for perhaps half an hour before taking a bit of a rest and trying to find a few flowers or insects to eat.

Males do this day after day, for months at a time, in the hope of attracting a mate. Clearly a lot of effort is expended in the process. If females wander near one of the display sites, the male goes into overdrive, displaying continuously, and leaving his display arena in a frantic attempt to court her. Other males are drawn to the commotion and may seek to get in on the act by displaying in the same area, thereby trespassing on the 'territory' of the others, in which case all manner of chasing and posturing results. But usually the females are not receptive and they get away as quickly as they can. After many hours of watching males display, I finally observed copulation. As soon as it was over, the female ran off.

During the first year I searched in vain for nests. Then, in the second year, I got lucky. By then I was so familiar with normal Houbara behaviour that I knew the moment I spotted the bird that it must be a female with a nest. The way she behaved was different from any I had seen before, and my eyes went straight to the point where I had seen her rise. There, to my intense joy, were three gorgeous bluish eggs (Plate 2). In my diary I wrote that they were the size of rugby balls, incredibly glossy and beautiful. Exaggeration, of course, in reality they average a little over 60mm long and 45mm wide, but in my excitement at finding them, they looked much bigger. A few days later I chanced upon another nest not too far from there, in a place where I could watch it from a distance. No males ever came near the nest. The one disappointment though, was that I was never able to watch a female with young. As soon as the eggs hatched, the female led the chicks away from the nesting area and I was unable to locate them.

Four

The Fuerteventura Chat

Although most of my time during 1980 and 1981 was devoted to studying the Houbara Bustard, I also spent some time studying a bird that is unique to Fuerteventura. In the years 1888 to 1891 the naturalist Edmund Meade-Waldo spent many months travelling in the Canary Islands, collecting and studying birds. During his time on Fuerteventura he noticed a small bird to which he was not immediately able to put a name. It was clearly a chat, but had quite different colouring to any he had seen before. He collected specimens and eggs and sent them back to England for examination. It proved to be a new species, the Fuerteventura Chat (Plate 4), a close relative of the Stonechat.

The male is a handsome bird, blackish above and white below with an orange splash across its breast, a prominent white stripe above the eye, and a small white flash in the wing. The female is altogether dowdier, a plain greyish brown above and pale below, somewhat resembling a flycatcher. The young bird is like the female but with a buff wash underneath contrasting with its white throat.

Fortunately enough, I found a few pairs nesting at the top end of the Fimapaire valley, just to the east of La Oliva, and easily within reach by bicycle. Every week or so, I would pedal to the valley for a break from studying bustards and a welcome change of scenery. I watched the chats building their nests in walls and under rocks, monitoring them to determine, amongst other things, that the eggs are laid at daily intervals and hatch after fourteen days, and that the young are fed in the nest for a little over two weeks.

Even within Fuerteventura, this bird has a limited distribution. It is fairly common in the *malpais* and in the deep valleys that run down the eastern side of the island, but is most numerous along the barrancos (dry river beds) that run down to the sea along the south and west coast. Fuerteventura chats live all along these barrancos, from the first shallow gully in the mountain side to the tamarisk thickets where the barranco meets the sea, where they can sometimes be seen flitting about on rocks on the beach. They are entirely absent from the open plains that cover much of the northern and central half of the island.

They seem to be remarkably sedentary, and this may explain why the species is absent from Lanzarote, which is just a few miles north of Fuerteventura. Despite its absence from Lanzarote, in 1913, when David Bannerman came to the eastern Canary Islands to collect birds and their eggs, he reported that his team had collected the chat from the islets of Montaña Clara and Alegranza, to the north of Lanzarote. His account is as follows:

> "Two days after our arrival on the island [of Montaña Clara] a remarkable discovery was made. Taking a stroll with his gun round the low ground, Bishop shot two Chats, which at the time I took to be the Fuerteventuran Chat – a sufficiently strange discovery even if this had been the case, as there are no Chats in Lanzarote or Graciosa. There were four birds together just behind the camp, but after the shot the others disappeared and were never seen again. Judge of our surprise to find that they were not examples of the Fuerteventuran Chat at all, but belonged to an entirely new and unnamed race...Whether the Chats were on migration is difficult to say; subsequently Bishop found them much more plentiful on Alegranza, but though I searched high and low for five more days [on Montaña Clara] I never saw a sign of these little birds again."

Given that Montaña Clara has an area of only about half a square mile, and that chats are relatively conspicuous little birds, this is odd indeed. Even more surprising is that no chats were ever seen on either of these islets again. Neither is it quite clear why Bannerman thought the birds that were collected on Montaña Clara were a different race to the Fuerteventuran birds. I visited the British Museum to examine the skins of the chats he collected on Fuerteventura and Montaña Clara, and it is hard to see any real difference between them. However, it would seem that the chat did occur in the northern islets, and this being the case it must once have occurred on the intervening island of Lanzarote. Perhaps its range expands and contracts with variations in the climate.

An unfortunate complication that arises from its erstwhile existence on the islets north of Lanzarote is the tedious matter of what to call the bird. Today it is endemic to Fuerteventura, and the name Fuerteventura Chat seems entirely appropriate. However, this does not stop the purists from pointing out that since the chat has once been found on another island, it needs a more encompassing name: hence it is often referred to as the Canary Islands Chat. Then there is its close relationship to the Stonechat to consider, hence the official name of Canary Islands Stonechat. For some reason, despite the fact that all birds have a unique scientific name – the chat in question is *Saxicola dacotiae* – the ornithological world also wants all birds to have a systematically derived common name that we should all use. Personally, I find the name Canary Islands Stonechat too much of a mouthful and simply call it the Fuerteventura Chat. Happily, I am not alone: BirdLife International's 2002 Conservation Action Plan for the species uses the same name.

In view of its rarity and isolation, and concerns for the survival of the Fuerteventura Chat, studies have been carried out to estimate the population. There are thought to be at least 750 pairs, and it seems to be doing quite well. I certainly haven't noticed any decline during my visits to the island over the last thirty years. No need then, really, for a conservation action plan.

FIVE

An idea becomes a reality

Some time during my stay in La Oliva, I noticed an old mule track running south across the plains. It was this track that first stirred the thought of a walk from La Oliva to the southern end of the island. I wanted to set off with my rucksack and just carry on south in a carefree way, but I knew that I had to keep going with the Houbara studies, even though I was finding it really hard. Nevertheless, an idea had been born.

Once the Houbara field work was completed, I returned to England to write up my thesis. It was a couple of years before I visited the island again. When I did it was like revisiting an old friend, and I have found one reason or another for going back every few years since. It was many years, though, before the idea of a walk down the island resurfaced. Some time in the mid-1990s I purchased a set of the military maps of Fuerteventura, and took to spreading them out on the floor. I wondered about possible camping places, how to link them together on a single walk, and how long I would need to complete the journey. Would two weeks be enough or did I need three? If I wanted to do the walk at a sensible pace, not too hurriedly, and with enough time to rest here and there, it always came to three weeks. How would I find enough time?

Then, towards the end of 2007, at the age of fifty, an opportunity arose to take early retirement from my job in the civil service. After weighing up the pros and cons I decided to take it and become an independent environmental consultant. The lump sum payment I would get as part of the early retirement package would allow me the freedom to stop work for a few weeks before doing so. If I really wanted to turn my dream into reality, I realised that there would never be a better time to do so. The question that remained was whether I really wanted to do it or not. The fact was that I had never undertaken a solo walk of more than a single day, and I had never walked more than a few miles with a rucksack. For me, therefore, this would be a real challenge, especially as I was proposing to scale several of the islands' highest peaks along the way. Was the whole idea a foolish whimsy?

To add to my uncertainty, I had booked a cross-country skiing holiday in Finnish Lapland with my daughters Clare and Katy. I had never been skiing before, and had some doubts as to whether I would come back in one piece! The skiing holiday was in mid-February 2008, and the slot I had for my walk in Fuerteventura was in March. I would let fate decide: if I came back from Finland in a fit state I would do the walk. We had a wonderful time in the snowy forests of Lapland. As well as cross-country skiing we went on a husky ride, we tobogganed down an impossibly long hill, and we even saw the aurora borealis. And after all the skiing I came back fitter than ever. I had no excuses left: I had to do the walk.

In twenty days, and with lots of help from my wife Eiluned, I purchased the equipment I needed, including rucksack, lightweight tent and sleeping bag, booked a reasonably priced flight and decided on a rough itinerary for a three-week venture. To get fitter I ran five miles every other day, and I walked long distances with the rucksack filled with my new equipment. To my relief I found that I could cope with the weight.

For anybody intent on walking in Fuerteventura, two items that I took with me are worth a mention. The first is the map that I used, which was the AA 1:50,000 scale map in the Island series. This was superb. So good in fact that although I also had the military maps with me I rarely referred to them, and indeed could have managed quite well without. The second item was a dog dazer, which I took to ward off any seriously nasty dogs that I might encounter along the way. I never used it, not because there were no large, fierce dogs; the island is as full of them now as it always has been. It was simply that every unpleasant dog was securely tethered, and this seems to be a rule that is now strictly adhered to all over the island.

Thus prepared, I set off for Fuerteventura, at the south-western extreme of the European Union, having been to the extreme opposite end of the union just three weeks before.

★

Instead of going directly to Fuerteventura, I wanted to fly to Lanzarote and then cross to Fuerteventura on the ferry. This would have a number of advantages. Firstly, there are more flights from the UK to Lanzarote than there are to Fuerteventura, and I was able to book a flight from Stansted, which is the nearest airport to my home on the Cambridge/Suffolk border. Secondly, I would arrive on the island at Corralejo, which would be the starting point for my walk. Thirdly, and most importantly, it seemed to me that arrival by ferry would have more of a sense of moment. I could watch the island growing in front of me, and there would be none of the tedium of customs and baggage reclaim, just an uninhibited walk from boat to shore.

I arrived at Arrecife airport on Sunday 9th March 2008, and took a taxi to Playa Bianca, on the south coast of Lanzarote. From my hotel that night I could see the lights of Corralejo, and indeed a rather spectacular firework display, across the Straights of Bocaina on the northern shore of Fuerteventura. The following morning I walked the few hundred yards to the ferry terminal. The choice was between the superfast Fred Olsen Express, which takes about a quarter of an hour to do the crossing, and the slower Armas ferry which takes a little over half an hour. I chose the latter on two accounts. Firstly, the slower ship allows better views of any seabirds, and secondly because the more leisurely approach would allow me to drink in the approach to the island.

I love the view of Fuerteventura from Lanzarote, and today I enjoyed it from the deck of the *Volcan de Tindaya* with a mix of emotions. For me, returning to the island is always like returning to a long-lost love, but on this occasion there was also a tinge of fear and apprehension. I was about to embark on the most physically challenging thing I had done in my life, and I wasn't sure I would be up to it. As Corralejo slowly drew nearer, I tried to work out from the map which of the mountains was which. I concluded that the peak dominating the left-hand end of the island (the eastern end) was Escanfraga (533m), and that the pyramid in the middle distance was Muda (689m), the highest mountain in the north of the island which I hoped to climb on the third day of the walk. In the foreground were the volcanic cones that I would be walking amongst later that day.

About halfway across the strait I saw my first Cory's Shearwaters of the trip: two birds gliding just above the waves in typical, stiff-winged shearwater fashion, showing alternately the brown upper side and white underside as they flew first with one wing almost touching the water and then the other. The shearwaters breed in their thousands on the islet of Lobos, which the ferry passes just before arriving at Corralejo. They are a common sight off Fuerteventura between late February and November. They only visit their nests at night, and large numbers congregate just offshore in the evening to await nightfall.

Although I saw none on this occasion, flying fish are also often seen from the ferry. These extraordinary fish, about 30cm long, can glide up to about 100m just above the surface of the sea, using their extended pectoral fins as stiff wings. They can be seen with the naked eye, but more easily by scanning the water with binoculars. It always comes as a great surprise when the first one appears then disappears again into the waves, because the natural assumption is that anything that flies must be a bird.

★

The ferry slid past the volcanic cone of Lobos and docked at Corralejo quay, at the north east tip of Fuerteventura. I shouldered my pack purposefully and perhaps with a little pride, and strode down the gang plank. Not surprisingly, the foot passengers all turned left towards land, but I turned right to start my walk at the very end of the island

– which I judged to be the far end of the quay. Here, a man and his young son were fishing and a black cat was lying in the warm sun watching them, more in hope than expectation I imagined.

The urge to start walking was very strong, but it seemed important to take in my surroundings first. I wanted this to be more than just a walk. So I took off my pack, lent it against the wall at the end of the quay, and forced myself to sit and watch. Another fisherman arrived, and then I noticed that there were dozens of fish in a loose shoal.

The waters around Fuerteventura are particularly rich in fish: around 400 species have been recorded, and the harbour at Corralejo is a good place to catch a glimpse of the rich marine life that lives in the area. One of the reasons for the rich marine life is the extent of shallow water around the island, particularly between Corralejo and Lobos. Perhaps the most important fish of all is *La Vieja*, literally 'The Old Woman', which lives down to a depth of about 50m. In appearance it is an exotic mixture of maroon and purple with a splash of yellow, and it can grow to half a metre in length and weigh up to 3kg. It is delicious to eat and is on the menu of all the best restaurants on the island.

In the past, the abundant fish were hunted by the Monk Seals that had a colony on Lobos. In fact, the island is named after them, as *lobo marino* is Spanish for seal (or literally 'sea wolf'). Unfortunately, the seals were hunted to extinction some time in the late nineteenth century, and today they are very rarely seen.

★

I turned round to find that not one but two black cats were taking a great interest in my pack. Perhaps they could smell something edible in it and were hoping to be fed. If so they were unlucky, but I hoped two black cats were a good omen!

From the quay I had the choice of walking left along the main promenade into town, or right along the residential north shore. I chose the latter. Although a little shabby, the 'Paseo Martino Bristol' is a pleasant walk, and the rocky beach always has a few shore birds (Box 5). Big lava boulders have been placed beside the promenade to keep the sea at bay. Growing amongst them, and on the upper shore beyond I noticed two typical seashore plants: Sea Rocket *Cakile maritima* with its spikes of delicate pale lilac flowers, and *Zygophyllum fontanesii*. The latter is a curious-looking plant known as Sea Grape on account of its swollen, often yellowish leaves that look like small fruits. It is the most characteristic seashore plant on the island.

About halfway along the north promenade, at the end of Calle de la Caleta, there is a solidly built structure consisting of a circular mound perhaps 20m across and several metres high surrounded by a wall of black lava stone. It is linked to a second, smaller, structure by a rather delicate-looking stone bridge. At some stage the mound was levelled to provide a crude platform for an unattractive stone bench, and tamarisks were planted on it. There is nothing to say what it is, and I can find no reference to it in any book. Perhaps, like so many of the ruins on the shore around the island, it was once a calcinating oven. Lime was a key export from Fuerteventura at one time. But here at the most northerly point in Fuerteventura, overlooking the narrow straight that separates the island from Lanzarote, could it have once had a defensive purpose? A curious feature is that the walls of the main mound were constructed in two phases. The lower section has a deeply wavy top, and this has then been levelled by infilling with a second layer. This appears to have been done for artistic reasons and to make it into a viewing point.

From there I headed towards the centre of town, passing the Plaza de la Molina de Juan Morera, with its small windmill. The mill and square were renovated in 2005. This is a good example of the type of mill constructed in the nineteenth century with the sails supported by a small structure rather than the more traditional circular stone buildings. The near constant wind makes Fuerteventura a good place for windmills, and there are still many to be seen throughout the island.

★

Originally, Corralejo was a small fishing village, but even in the late 1960s, John Mercer, in his fine book about the island (by far the best in English) described it as a building site. By then the glorious sandy beaches had already been noted by developers, and the first few hotels and apartment blocks were being built. During my time in La Oliva in the early 1980s, Corralejo was the bright lights, the place we went to for a night out. At that time fishing was still important, and beyond the main square the streets were dimly lit and a little mysterious.

Over the last three decades Corralejo has been transformed into a major tourist centre owing to the wonderful sandy beach immediately to the south. The sandy area immediately inland from the beach is designated as a Natural Park (Box 6). The main street, Avenida Nuestra Señora del Carmen, is now one long line of shops, banks, restaurants and bars, all with a rather British feel to it. A road train runs up and down the street, carrying both newly arrived, white-skinned tourists and the glowing pink tourists who have been around for several days.

★

Before leaving Corralejo I stocked up with a few basic provisions, including two litres of water, which I hoped would be enough to keep me going until I reached La Oliva the following day. Then I retired to a café for some final refreshments before setting out in earnest, and turned my attention to the stowing away of my provisions. Unfortunately, it was physically impossible to get the provisions in my already full rucksack! In planning for the trip, I had been stupid enough to think about the weight of water that I would need to carry, but not about its volume. Although I didn't want to admit it to myself, it was now obvious that the rucksack was too small. After a brief moment of panic, I realised that I could put some of the food in the very small day-sack that I had brought with me.

With one final push I forced the water and some of the more robust provisions into the rucksack, which now bulged at the seams. The rest went easily enough into the day-sack. Without realising it, I had, of course, been providing entertainment for the other people in the café. I heard an English woman behind me wondering out loud what I would do if I needed to get to something at the bottom. This was a very good point, and one that was not successfully answered for the remainder of the walk.

The lady in question and her husband were from the Midlands. As I sipped my coffee I told them about the walk I was about to embark on. It was good to have someone to talk to. Having finished my coffee and packed as best I could, I walked out of the café. This was it: my journey was about to begin in earnest, and now I would find out whether I was up to the challenge I had set myself. The pack was so heavy I could hardly lift it high enough to get my arms into the straps, and there was a rather unattractive struggle on the pavement outside the café before I managed to do so. I hoped that my Midlands friends were not watching because it would be clear to them that I was a novice at this game, and they might not rate my chance of success very highly. There was nothing for it but to carry the small bag in my hand. Thus loaded, I set off determinedly but a little apprehensively up the street. I hoped I looked more confident and capable than I felt. The enormity of the task that I now faced weighed heavily on me. I was definitely throwing myself in at the deep end.

The road on the southern outskirts of Corralejo is now lined with young palm trees. The fashion for planting avenues of palms along the entrances to the towns and villages started in the late 1980s, and many now have them. This has certainly transformed the approach: in the past one went straight from desert to village with no introductory statement. I have no doubt that many people prefer to have a palm-fringed entrance to town, but for me it is a prettying-up exercise that detracts from the character of the place. A somewhat surprising feature along the entrance to Corralejo is the replica pirate ship that towers over the road and is the centrepiece of the new Baku leisure park with its large pool, flumes and other attractions.

Eventually, I reached the sports stadium from where I was expecting to find a path to Lajares, but the ring road has now been extended, and there is a roundabout with a rather fine metal sculpture (Plate 24). The *Estrella del Norte* (Star of the North), is a small sailing boat set on a wavy blue and white base that mimics the sea. Five cut-out shapes in the sail depict the sports of swimming, basketball, volleyball, surfing and Lucha Canaria.

Lucha Canaria is a form of wrestling that may or may not have its origins in the wrestling performed by the pre-Conquest inhabitants. Either way, the modern version of the sport is something that the Canary Islanders are proud of. During my time in La Oliva in the early 1980s, I could sit on my flat roof and through binoculars watch bouts of Lucha Canaria taking place on an open space at the back of the village. The object is to throw your opponent to the ground or to make any part of him other than his feet touch the ground. Today it is played as a team game, with twelve on each side.

Inside the boat sculpture a fisherman points meaningfully towards the North Star, which from here is just to the left of Lanzarote. The North Star is, of course, important for navigation at sea, but north is also of great symbolic importance in Fuerteventura. It is from the north that the first European explorers came in the fourteenth century, to be followed by conquerors a century later. In the second half of the twentieth century, it was from the north that the tourists came and changed the island forever. Change has always come from the same direction in this island.

North or north-east is also the direction of the wind on all but a few days of the year. The wind is generally quite strong, on average seventeen kilometres per hour, often more in summer. The near-constant wind makes the island a perfect place for windsurfing, for which it has an international reputation. The north wind was also an important reason for choosing to walk the island from north to south. On the few occasions when a twist in the route required a brief period of walking north, the wisdom of this decision was very apparent. Also, because I had lived in the northern part of the island, it seemed natural to start at that end.

Box 5: Birds of the shore

The most conspicuous bird on the seashores around the island is the Yellow-legged Gull, a bird very like the Herring Gull of northern Europe, but with somewhat darker wings and back. Although not particularly numerous, pure white Little Egrets are also often seen, whilst Grey Herons occur in very small numbers.

Smaller wading birds can be seen on the shore throughout the year, but are more abundant in winter. Of these, only the Kentish Plover nests here, but Ringed and Grey Plovers, Whimbrel, Redshank, Greenshank, Turnstone, Sanderling and Common Sandpiper are also frequent.

Box 6: Corralejo Dunes

South-east of the town is the Corralejo Dunes Natural Park, designated for its rare flora and fauna. It is eight kilometres from north to south and two and a half kilometres across at its widest point, and is one of the largest dune systems on the island.

The most spectacular approach to the dune system is from the south. The first part of the road north from Puerto del Rosario is unattractive, but beyond the rose-tinted hill of Montaña Rojo, the glorious white dunes of the natural park come into view. The sand is made entirely of shells, hence its colour. From time to time the wind drives the sand across the road and bulldozers are required to keep the road open. One is reminded of blizzards and snowploughs.

Head inland from the road for a few minutes into the dazzling white wilderness and you are transported into another world. Sight and sound of civilisation are soon lost, but do be aware that naturists like this place too!

Naturalists, however, would do better to enter from the north and keep to areas of partially consolidated sand near the western part of the park. This is where the special flora is to be found, and where Houbara Bustards can still be seen – hence the rather comical natural park signs in the form of a cut-out Houbara. The little shrub Yellow-flowered Restharrow *Ononis hesperia* is common, together with a small blue-grey succulent *Polycarpaea nivea*. Another yellow flowered plant that is fairly common in all parts of the island is a birdsfoot trefoil *Lotus lancerottensis* (Plate 12). Rarer than this, though, is the white-flowered African sand lily *Androcymbium psammophilum*. The plants here are of a subspecies found only in sandy areas in northern Fuerteventura and on Lanzarote. It has rosettes of long, strappy, pointed leaves forming green stars over the sand. The white flowers are produced in early spring. Even more striking is the broomrape *Cistanche phelypaea*, the chunky yellow flower spikes of which appear after spring rain. This plant is found throughout the Mediterranean, and is parasitic on the saltwort shrubs that are so abundant here.

Six

Malpais

The aim of the first day's walk was to follow the eight-kilometre track from Corralejo through the extraordinary landscape of volcanoes and *malpais* to the village of Lajares. From there I would continue south and make my first camp in the sandy wastes beyond. However, finding the path to Lajares proved to be much more difficult than I had imagined. From the Lajares end, a good path has recently been constructed and can easily be found. However, the Corralejo end proved to be less straightforward, and neither of the tracks leading away from the Estrella del Norte roundabout looked promising. One led to a quarry and the other towards what appeared to be a building site. A track leading down past a house with a fine garden and a rather extraordinary volcanic vent seemed to be going in the right direction, so I took it. I stopped in the gateway to take a photograph of the house and garden but suddenly a huge dog I had not noticed leapt for me. Thankfully it was tethered, and its leap fell just short, otherwise the walk might have ended before it had really got started.

Thanks to a helpful man who appeared from nowhere, I finally found my way to the right track. It was gone noon by the time I finally escaped from the clutches of Corralejo and cresting a steep rise, looked out across an unspoilt volcanic landscape. It felt as though the walk had started at last. I took off my pack and celebrated with a quick snack and a drink of water. It now occurred to me that I could fix the little day sack I had been carrying to the back of my rucksack by tying it on with my spare shoelaces. This achieved, I set off with renewed confidence.

★

The track to Lajares crosses a vast area of lava known as the Malpais de Buyayo, winding its way between a series of volcanic cones, the most recent of which are just a few thousand years old. The Buyayo crater itself dominates the skyline south of Corralejo. Although this rocky landscape is very young in geological terms, there has been enough time for erosion to allow some vegetation to grow. This is in contrast to the very recent volcanic area of western Lanzarote, where the substantial eruptions in 1730-36 have created what remains an utterly barren, black landscape. In contrast, the Fuerteventuran *malpais* tend to look greyish-green from a distance because of the rich growth of lichens (Box 7).

I was pleased to see that the side of the track was dotted with flowers, which meant that there had been good spring rains. As I followed the track south into the increasingly wild landscape, I stopped frequently to enjoy the flowers and to note the variety of species: all of them common on the island, but lovely to see nevertheless. The more obvious species here included the vivid blue Scarlet Pimpernel *Anagalis arvensis* (a contradiction in terms as the variety that grows in Fuerteventura is blue rather than red), the little yellow-flowered annual marigold *Calendula arvensis* and white flower spikes of the small Hollow-leaved Asphodel *Asphodelus fistulosus*. Here and there were also low clumps of the mauve stock *Matthiola fruticulosa*.

After rising to an altitude of around 100 metres, the track levelled out. To the left was the steep cone of Buyayo, and to the right was a flatter area from which sprang numerous volcanic humps and a high pyramidal mound of lava. This pyramid was topped by a large, jagged rock from which a Buzzard watched me. For some reason, in this setting, it seemed surprisingly menacing.

Buzzards were very scarce until quite recently. From early ornithological accounts, they appear always to have been uncommon, probably reflecting the relative rarity of rabbits on the island. Happily though, they are now much more numerous. Buzzards seem to have benefited from the Barbary Ground Squirrel (Plate 40), which was introduced in the south in 1965 but has now become abundant everywhere. This conclusion is supported by the fact that the Buzzard has become extinct as a breeding bird in Lanzarote, where the ground squirrel is absent.

The ground squirrel is a native of Morocco and Algeria, especially the Atlas Mountains. It was introduced from the African coast opposite Fuerteventura, which was a Spanish territory until the 1970s. Despite its terrestrial habits it is a true squirrel, measuring about 30-40cm long, almost half of which is the bushy tail. It lives in walls and amongst rocks and is regarded as a pest by farmers as its diet includes grain and fruits. On the plus side, it is something of a tourist attraction, and there are places where this rather appealing stripy rodent takes food from the hand. Away from the tourist hotspots, it is wary and might pass unnoticed, but its somewhat wheezy, curlew-like whistles are heard on every hillside. I could hear one calling from a jagged pile of rocks. No doubt the Buzzard could too.

★

I stopped for lunch overlooking a slight hollow where there were saltwort bushes and a scatter of colourful flowers. I removed the heavy pack gratefully and settled out of the wind behind rocks crusted with grey lichen. Away to my right was the sea with the distant hills of Lanzarote in dappled sunlight beyond. A migrant Willow Warbler, pausing here for a few days on its way north to Europe, flitted about in the saltwort. There was also a family of Fuerteventura Chats. They were curious and approached to within a few metres, the male singing from a rock behind me.

★

The track continued through the volcanic landscape, threading its way between a number of cinder cones. At one point I veered off the main track to follow a long abandoned mule track from which there was a fine view across the *malpais* to the north coast. A handsome male

Trumpeter Finch was gorging itself on the ripe red lycium berries. The Trumpeter Finch is a perky, sandy-coloured little desert bird. The handsome male is rather stockily built, with a stout coral red bill and pink flushes on breast, rump and wings. Its most extraordinary characteristic is its song, a bizarre buzzing that does sound surprisingly like a toy trumpet and is frankly comical. The Trumpeter Finch is almost always in small flocks, and is to be found anywhere where weeds have gone to seed, from the edge of town to the tops of the highest mountains. Like all seed-eating birds it needs access to water, and will travel far to drink.

Rejoining the main track there was an isolated scatter of low buildings set amongst a patchwork of rocks and fields. After the stark rocky wastes of the *malpais* the green looked impossibly vivid, a patch of hope in a sea of despair. The winter rains had produced a flush of grass, which would later be collected as fodder for the livestock. In the recent past the buildings would have been the mean dwellings of hardy goatherds. Now, such a way of life has all but gone: tourism brings real money and a much easier life, but is it more satisfying? The rock walls that had been so carefully built to keep the goats out of the crops were now broken down and useless.

★

Here and there amongst the usual dull green saltwort bushes were jauntier splashes of yellowish green. These were the leaves and flowers of the succulent shrubby spurge *Euphorbia regis-jubae* (Box 8). Growing amongst the Euphorbia was another succulent shrub with bare bulbous 'fingers' sticking up from the stem. This was *Kleinia neriifolia*, which sprouts sparing clusters of leathery, dull green leaves after rain, and rather surprisingly, ragwort-like flowers: it is indeed a succulent ragwort (Plate 8).

In one place there were two very different pink legumes: the tiny birdsfoot trefoil *Lotus glinoides*, which also occurs in North Africa, and the much showier restharrow *Ononis serrata*, a Mediterranean species with masses of pale pink flowers 10mm across. Two other plants growing here that I was to meet with frequently along the way were a small dock with red fruits *Rumex vesicarius* and an equally small yellow flowered ragwort *Senecio glaucus*, another Mediterranean species.

★

Around mid-afternoon the rain clouds that I had noticed earlier over Lanzarote were now almost upon me. Fortunately, I knew that shelter was close to hand. I had now come to the far end of the rather eccentric footpath that has been made for tourists to walk from Lajares to the impressive crater at Calderón Hondo (Deep Crater). This is a ribbon of neatly laid volcanic rocks that is almost impossibly uncomfortable to walk on, even in rugged walking boots. On the northern slope of the volcano it also leads to a reconstructed goatherd's hut on a low ridge. I could see the hut now on the hill above me, so I left the main track and followed the path upwards. It is a simple two-celled structure of rough volcanic rocks, a separate entrance to each cell on the sheltered southern side, complete with traditional flat mud and straw roof. I was glad to reach it as there was now a stiff northerly wind and a definite threat of rain. Inside there were just bare floors and enough room to lie down, but the ceiling was too low for me to stand upright. Of course, it is unused today, but I doubt there would have been much in the way of furnishings even when it was in use. I find dwellings such as this immensely humbling.

Once inside the hut, I fished out my waterproofs to make them more readily available in the event of a downpour, and put the waterproof covering over my rucksack. The last thing I wanted was to get my camping equipment wet. In the event there was a faint dampness in the air for a few minutes, but it didn't rain. Being British I tend to equate heavy low clouds and strong wind with an imminent downpour, but in this desert climate, dark clouds often mean nothing and I have learnt that they are generally best ignored.

★

In my first year of bustard studies I assumed that Fuerteventura never experienced heavy rain. This was not an unreasonable assumption given that the island was clearly a desert, and I had never witnessed rain there. It was, however, quite wrong. Just occasionally, perhaps once or twice a year, it does pour. One day early in 1980, I decided to camp overnight near Tindaya, where I had been watching a group of bustards. I walked the eight kilometres from La Oliva in late afternoon, with a view to spending the night in a roofless building I had noticed nearby. I would then be able to watch the bustards at dawn. I put my sleeping bag and some provisions (but no waterproofs) in my rucksack and headed off down an old mule track that heads directly across the plains to this area. (This was, in fact, the mule track that had inspired me to undertake the walk I was now undertaking.)

Almost as soon as I had reached the roofless building in question, it began to drizzle. I sheltered as best I could behind its walls and waited for it to stop. It didn't! Out to sea the sky grew inky black, and I now noticed that the wind was coming from the west, something it had never done in the preceding weeks. Before long it was raining rather than drizzling and, more ominously, there was the sound of distant thunder. On top of that, night was beginning to fall. At some point, with lightning flashing all around me, and being utterly drenched, I realised that my situation was hopeless. There was nothing for it but to head home. I now learnt another thing about Fuerteventura. The dusty soil that covers the plains turns to a sticky clay when it rains, and it clung to my shoes in great clarts making the going very heavy indeed.

It was the kind of rain normally associated with rainforests. Rather than endure an eight-kilometre walk through the drenching rain I headed for the nearest road in the hope of getting a lift. Even so, it was an awful four kilometres to the road, first across the plain then up through the dimly lit streets of Tindaya village. When I finally reached the road, one of the first cars stopped. I was worried that because I was soaked to the skin and very muddy, they would drive on. But no, they beckoned me in. I was so grateful.

★

Once it was clear that it was not going to rain, I left the goatherd's hut, and followed the path up to the rim of the Calderón Hondo crater. Beside the path were the little pink flowers of Greater Sea-spurrey *Spergularia media*. At the top there is a wooden platform from which you look out across the spectacular 100m-deep crater, to the opposite cliff with its wavy layers of solidified lava. It would be worth the climb simply for the wonderful views to the north even if the crater was not there. In the foreground the rocky *malpais* is frosted with pale grey lichen, relieved here and there by patches of brown and green where corrals have been constructed from the lumps of ejected lava that cover everything here. In the middle distance were the scattered white dwellings I had passed earlier, set amongst the oasis of green, and beyond them two isolated volcanic cones (Montaña Lomo Blanco and Montaña de la Mancha) with the grey sea and the outline of Lanzarote in the far distance. As on my previous visit, there was a fierce wind at the crater rim, so perhaps this is a regular phenomenon caused by the geography of the site. It certainly adds to the atmosphere of the place, and my hat would have been lost in seconds if it hadn't been securely attached. There must be a good assortment of hats at the bottom of that crater.

Rejoining the main track, I pressed on south. In contrast to the Corralejo end where I had seen only an occasional farmer's truck and a lone cyclist, now that I had rejoined the tourist trail I passed quite a few people heading from Lajares up the hill to the Calderón Hondo crater. This is a well-used walk, and rightly so.

Beyond a low rise, glorious views to the south opened out in front of me, a view that encompassed much of the next two days walking. Below me was the village of Lajares sprawling rather untidily over the plain. I found a sheltered rock just below the track from where I could admire the view and take stock of the journey in front of me. Immediately to the south was the striking cone of Montaña Arena: at 421m, much larger than the volcanoes I had been walking amongst. La Oliva, which I would be walking through tomorrow, lay on the other side of it. Beyond, dominating the skyline, was the pyramidal peak of Muda, the highest mountain in the northern part of the island. If everything went to plan, I would climb it the day after tomorrow. Further to the right, rain clouds hung over the Betancuria mountains in the far distance, where I hoped I would be in a week or so. Nearer to hand, on the far right-hand side of my view, and stretching out beyond Lajares, was the sandy waste where I hoped to make my first camp.

Now that I was nearing the end of the first day, I was starting to feel apprehensive. Was this venture a good idea or had I taken on something too big? This would be the first of many nights when I would need to find somewhere suitable to pitch the tent. At least tonight I would be camping in an area that I knew: the Lajares sands marked the far end of my Houbara study area. Anyway, there were three and a half hours to go before dark, and that would leave time enough for a well-earned meal in Lajares.

★

At its southern end the track drops steadily down towards Lajares, skirting the reddish-coloured cinder cone of Montaña Colorada, the coloured mountain. It must once have been beautiful but, like so many cinder cones in these islands, it has great gouges in it where it has been quarried. Its perfect natural form has been ruined for ever. The cinder, locally known as *picón*, is spread over sown crops to help stop them drying out, and also, by increasing the surface area, to increase the watering effect of dew. Although local mining for *picón* has largely ceased, huge quantities of it are now mined commercially to supply the construction industry, with serious consequences for the landscape, as I would see later on during the walk.

A little before the road, the track turned to the left, and the peak of Muda was now straight ahead of me. At that moment I felt a wonderful sense of adventure.

Box 7 Lichens

In many parts of the world, lichens are associated with places that receive high rainfall, but there are many desert lichens. Fuerteventura has a rich lichen flora of about 200 species. Although rainfall is low, the frequent heavy dew on the island is a significant benefit. Another factor that helps make Fuerteventura a good place for lichens is the purity of the air, since many species are very sensitive to air pollution.

Particularly noticeable in the *malpais* of Fuerteventura are lichens of the genus *Ramalina*. One of the most abundant and characteristic is *Ramalina bourgeana*, which appears to be confined to the Macaronesian Islands (Madeira, the Azores and Canary Islands). It has rather long pale green straps that make a substantial contribution to the overall colour of the *malpais*. Almost every rock on the *malpais* is covered with a mosaic of different coloured lichens, often greenish or grey, but splashed here and there with the vivid orange of *Xanthoria* and *Caloplaca* species.

Lichens of the genus *Rocella* were one of the incentives for European colonisation of the islands in the fifteenth century. Early visitors were quick to note that this lichen, commonly known as orchil, grew in abundance on cliffs and rock faces. It grows with other lichens on the rock faces of the *malpais*, but more abundantly in the mountains. Orchil was an important source of red and purple dye in mediaeval times, and was certainly an important part of the economy of the island in the fifteenth and sixteenth centuries. No doubt orchil growing in easily accessible places was quickly collected and exported, and brave work would have been required to collect it from high cliffs. After an initial glut, the plant would have become scarcer, as lichens do not grow very fast. However, it was still collected until the nineteenth century, when synthetic alternatives became available.

Box 8   *Euphorbias and Echiums*

There are no less than twelve species of succulent spurges (Euphorbias) in the Canary Islands, most of which are endemic to the islands. Four occur in Fuerteventura, of which *Euphorbia regis-jubae* is by far the commonest. The other species are the larger *E. balsamifera,* the organ pipe-like *E. canariensis* and the spiny *E. handiensis*, which is endemic to Fuerteventura.

Euphorbia regis-jubae favours slightly moister sites such as gullies, where it sometimes forms dense stands up to a metre or so high. The main stem and branches are thick succulent cylinders, and the plant is topped with a mass of bright green, rather leathery leaves and yellowish flowers. It is named, rather fittingly, after King Juba II, who was the king of Mauretania (which included present day Morocco) from 29 BC to about 20 AD during the period of Roman domination. He mounted an expedition to the islands, which left an account of the succulent plants found growing there. However, whilst the species is named after the king, the whole genus is named after his physician Euphorbus. That is quite an honour as the genus is enormous. There are around 2,000 species worldwide, exhibiting as much variety of form as any genus of flowering plants on earth.

Another group of plants that the Canary Islands are famous for, at least amongst gardeners, is the genus *Echium* of which our own Viper's Bugloss is a member. No less than twenty-five are endemic to the islands, including spectacular plants such as *Echium wildpretii* in the high mountains of Tenerife and La Palma, which produces spikes of red flowers two metres high or more, and the extremely rare blue-flowered *Echium pininana* of La Palma that grows to a staggering four metres and is often planted in exotic gardens in England. Only three species are native to Fuerteventura, the commonest of which is the very small *Echium bonnetii*, the vivid purplish-blue flowers of which are amongst the most frequently encountered on the island.

SEVEN

First camp

It was late afternoon when I reached the outskirts of Lajares. In the five hours since I had set off from the bar in Corralejo I had covered about eleven kilometres. The pace was slow, in part because I needed to stop from time to time to take off the heavy pack, but also because I was determined to see as much as possible on the walk, not just to walk for the sake of walking. I was quite tired, though that was hardly surprising, but what did concern me was that the sole of my left foot was quite sore. A small blister was forming, a worrying development on day one of a long walk. Had I chosen the wrong footwear? If so, there was nothing that could be done now to remedy the situation. The possibility of complete failure suddenly confronted me. Pushing that thought to the back of my mind as much as I could, I put a plaster on in the hope of preventing further deterioration, and walked gingerly on into Lajares in search of somewhere to eat.

In the not very distant past, Lajares was a simple village with a shop at the junction in the centre and a couple of old windmills near the plain little church of San Antonio. In recent years the village has prospered. Rather bizarrely, given that it is quite some distance from the sea, it has become something of a centre for surfers and there are several shops selling surfing gear. As a consequence, it now has a thoroughly international community. There are smart new houses going up everywhere, including some rather luxurious holiday villas, and there is an air of prosperity.

A number of cafés/restaurants have also appeared but I was disappointed to find that most were shut. It was probably the wrong time of day. The only bar that was open had basic snacks on offer but nothing substantial. I would have to make do with camomile tea and a cake, which was rather a depressing thought. It was certainly not the indulgent meal I had had in mind.

Not much refreshed, and without any great enthusiasm I headed out of the village towards the sandy plain that lies to the south of the noisy bypass. On the village outskirts was a fine old house that would once have been quite grand, but was now unwanted and broken. Nearly all the old houses on the island have been abandoned in favour of modern comforts. Typically, this one had big dressed stones at the corners, and a gaping roof of sagging timbers through which the low sun threw patterns of shade on its sturdy walls: a sculpture of textures and light that had once been a home. Surrounding the plot, one carefully built stone wall was still intact, another utterly broken with its lichen-crusted stones strewn over the ground, almost as haphazardly as when first ejected from the volcano millions of years before. In what was once the yard there were a few cacti, their stacks of green discs growing from the rocks as if placed by a sculptor. Emotionally, these ruins touch me. They stand as memorials to a way of life that has passed: a way of life that still hung on when I found myself living here in the early 1980s. In that sense, they are also a reminder of my own past and the lonely time I spent on the island when I was still a fresh-faced student.

★

Beyond the bypass is a shallow valley covered in sand, despite being seven kilometres from the sea. Low hills surround the valley on three sides, but it is open to the north, and sand has blown here from the northern shores, driven south by the predominant winds. In fact, marine sand has blown over much of the island over millions of years, but in most places the surface layers have transformed with the passage of time into a smooth calcareous crust. Here at Lajares, the crust is overlain by a thin cover of mobile sand. The flora is strikingly different to that on nearby rocks, and similar to that in the Corralejo Dunes Natural Park (Box 6).

This sandy plain is a desolate place that appears always to have been uninhabited: even the ubiquitous ruins are absent. For this reason I had earmarked it as an ideal, peaceful place to camp on my first night. A wide, sandy track runs through the centre of the plain, but I was surprised to find that there were lorries using it, going where I could not imagine, so I walked to one side of it. And then a big surprise, a huge hole in the ground right in my path! Where previously there had been nothing but scrub and sand, there was now a quarry tens of metres deep and hundreds of metres across. Vehicles operating down below were dwarfed by the scale of the hole. The quarry was clearly now being used as a rubbish dump, hence the lorries on the track. I therefore had to detour some way round to the left, edging between the rim of the quarry and the steep slope of the adjacent hill.

By the time I had completed my detour and was far enough from the hole to leave its noise behind, the sun was dipping rapidly towards the horizon and all warmth had gone from the day. I needed to find a place to camp, somewhere out of the keen northerly wind. Having considered several alternative spots I eventually decided on a slight hollow in the sand which was at least partly sheltered by a low mound and a few small *Launaea* shrubs. I knew there was an area with unusually large shrubs somewhere hereabouts that would have made an ideal camping spot, but it was nearly thirty years since I had been here, and I wasn't sure how long it would take to find it. I was tired and the light was fading rapidly, so this would have to do.

I pitched the tent for the first time and sat outside to eat my rather pitiful rations, looking out over the rapidly dimming valley of sand and scrub. I should probably have felt elated: my journey had begun and here I was in splendid isolation on the island I loved. And yet now that I was here I felt uncertain and lonely. As the light faded and the cool wind began to chill me, I heard the sweet, simple song of a Spectacled Warbler nearby. Then a second bird answered it, confirming its territory. The tent was close to a deep, scrubby gully that to them was home. This little bird is one of the few on the island with a tuneful song, and it sings strongly at dawn and dusk. It can be found anywhere with a decent patch of scrub, although it is quite secretive and easily overlooked, most often seen as a tiny brown ball dashing from one shrub to another, giving a harsh, scolding call as it does so. On this occasion its pleasing song failed to warm me.

It was almost dark by the time I retreated to the relative warmth and comfort of the tent. I was disheartened to find that the blister on my left foot had become much larger. Was the whole trip about to be ruined?

I had agreed with my wife that I would text her on my mobile every night to confirm my position and hopefully to reassure her that everything was OK. This would be my lifeline. As I was travelling alone it was sensible to make contact every day, although quite what she could have done if anything had gone seriously wrong is open to question. Now that darkness had fallen I turned the mobile on and composed a brief text, but failed to find a signal. This was a very lonely moment, my one connection with normal life had gone. But then the mobile found a signal and the text went. I now also found that I had an email from her, which gave me a big boost. I settled into my sleeping bag, which was warm and cosy, and simply lay there exploring my own thoughts and trying to settle my mind.

I was stirred from contemplation by the calls of a Stone-curlew flying low over the tent. Stone-curlews, rather like small bustards in appearance, are masters of camouflage. Their dappled and streaked plumage blends perfectly with stones and sand, and they sit quietly most of the day, being all but invisible. They become active at night, when their distinctive calls carry far across the plains. The very last light of day and the very first glow of light in the morning are special to them, and they seem incapable of allowing either moment to pass unannounced. They have a range of different calls, some clear and wistful, some hurried and rhythmic, and others that are rasping. The overall effect might be quite disconcerting if you didn't know what they were, but their calls were a great comfort to me now.

My thoughts returned to my situation, and I tried setting out the positives and the negatives. On the plus side, I had walked about seventeen kilometres and I was not overly tired. The distance I planned to walk tomorrow was only fourteen kilometres, and that was clearly within my capability. Carrying the heavy pack was certainly physically possible for me. One concern I had had was my left shoulder, which had been causing me pain on and off over the preceding months. I wasn't sure how it would react to the weight of my rucksack. I was surprised to realise that it was giving me no pain at all, and that was a real bonus. Also on the plus side, after two more nights under canvas, I could look forward to a night in a proper bed, because I was booked into a hotel in the resort of Fustes, halfway down the east coast. On the negative side, however, were the blistered foot and a sense of loneliness. In my experience, loneliness is never quite what it seems. It is often a close relative of, and perhaps the same thing as boredom, a need for company in the absence of other distractions. But in this case I think it was mainly caused by deep apprehension. One day out here might be a bit of fun, but what would three weeks of it be like? Would I be up to the challenge I had set myself?

But I told myself that I had to enjoy the walk. This was what I had dreamt of doing, and worrying my way through it just wasn't good enough. I had to get a grip. I now sought inspiration from my reading material. Because of the need to keep weight down, I had restricted myself to two smallish books. I thought long and hard about what to bring and decided on one book that would give me something to think about and another that would encourage me to meet the physical challenge. For the first need I brought *A Short History of the World* by H. G. Wells, which I hoped would be both readable and interesting. I would be able to use the time I had on the walk learning something. The other one was an extraordinary book of travel writing that had inspired me fifteen years before. This was *Wheelbarrow Across the Sahara*, by Geoffrey Howard. The reason for bringing this was that the hardships this crazy man endured, whilst pushing a Chinese sailing wheelbarrow from Beni Abbes in Algeria to Kano in Nigeria, were so extreme that they would clearly knock any minor inconveniences I would suffer on my modest walk into a cocked hat.

It was the 'wheelbarrow book' that now came out of the pack. Outside the tent, the sky was full of blazing stars and there was a slither of new moon. It was waxing and would provide light on future nights. Finally, before going to sleep, I played Radiohead's album *In Rainbows* on my Ipod. Somehow, I found it very comforting, and I got into the habit of playing it every night when I was under canvas. It will forever remind me of my walk in Fuerteventura.

EIGHT

Memories

My first night under canvas was on the cool side, but otherwise not uncomfortable. Despite my attempt to find a sheltered spot the wind caused the tent to flap quite alarmingly, but I soon got used to it. The Stone-curlews called on and off all through the night, and as usual, more intensely just before dawn. There were dogs barking in the distance, and not surprisingly I had a nasty dream about a dog! No matter where I camped on the island, there was always the sound of dogs barking.

I broke camp soon after dawn. The morning's walk would be on familiar territory, passing through my old bustard study area and then into my 'home town' of La Oliva for lunch. I was hoping to catch sight of at least one Houbara, for old-time's sake and to prove to myself that I could still spot them, so I told myself to go slowly and concentrate on what was around me: more easily said than done when you are carrying a heavy pack and there is the lure of proper food and good coffee.

To start with it was cool enough to need my fleece, but it quickly warmed up, and I had to stop to take it off. My foot was already really sore so I dug out the sandals and tied my boots to the outside of the rucksack. After donning sun cream and hat I set off again. As I did so two large birds suddenly exploded from the ground just in front of me, and another three rose a little way off. They were Black-bellied Sandgrouse, birds of dry steppe and desert edge that are particularly at home in sandy wastes. They are somewhat stocky birds about the size of a pigeon but with wonderfully marked plumage. The male is covered in orange-buff spots above, with a grey breast and head and a handsome rusty-yellow and black throat patch. The female is rather more subdued, more cryptically marked to make her even harder to spot when she crouches. Two characteristics of these birds make them very easy to identify. The first, as the name suggests, is the black belly patch that is conspicuous on both sexes in flight. The other is the remarkable call that I have always struggled to describe. However, during the skiing holiday in Lapland a few weeks before the walk, I heard almost the exact same sound. A ski-pole twanging under pressure sounds very like the call of this bird! It could perhaps be described as a liquid, wavering chortle. Once heard it is never forgotten.

★

A characteristic feature of the area I was now walking through, just east of the Lajares sands, is a series of deep gullies formed by the rain washing through the calcareous crust. The naturalist David Bannerman passed through here on his journey from La Oliva to Cotillo in 1913. At that time the track I was now looking for must have been the main pack route between the two villages. He wrote the following:

> "We passed over several dry watercourses, which were mere dry cracks in the earth, zigzagging to a barranco – these water-worn 'nullahs', though some twenty feet in width, with absolutely perpendicular sides, as if cut out with a spade, were about nine feet in depth. They were remarkably difficult to detect until one had almost stumbled into them, and would prove exceedingly dangerous obstacles when walking or riding at night."

In fact, the pack route avoided the biggest of the gullies, and good stone bridges are in place to cross them. Some of the gullies are a good deal more than nine feet deep. The largest one I crossed this morning was nearer twenty feet deep (6m) (Plate 27). It is no easy matter to cross these daunting scars, but somehow I did manage to find ways down and back out again.

★

One of the common flowers on the hillsides here, and indeed almost everywhere on the island, is a small yellow-flowered rockrose *Helianthemum canariense*. During my studies here in the early 1980s, I was intrigued to see a man quartering the hillside, evidently looking for something. What could he possibly be looking for? The answer came from John Mercer's book, in which there is a section on truffle hunting. There is apparently no direct connection between the truffles and the rockrose, but they grow in the same type of soil. John Mercer explains:

"A little rain and the fungus is at once to be found, becoming plentiful after ten days...Here and there amongst the *turmeras* [rock rose] the earth gets pushed up a little, as by a nascent mushroom, though nothing is visible. The hunter scrapes away and reveals a cracked greyish-brown lump, the size of a medium potato, though they can be as large as a fist...Cooked like potatoes, they are as good if not better, with their light flavour of smoked bacon."

★

By mid-morning I was walking on the old pack route. I was now amongst the dry hills at the western end of my old bustard study area. On the low ridge to my right were the ruins of Gavias del Carce, two abandoned dwellings on opposite sides of a small cultivated depression. When I was studying the bustards back in the 1980s I sometimes sat in these ruins to have lunch, wondering what life would have been like for the families that had lived out here in this harsh but beautiful environment. In some ways I would have loved to have experienced that life, though it must have been incredibly hard.

Once, when I had arrived at the more complete of the two houses, I could hear faint meowing coming from inside. At one end there was a closed door, and the cat was inside. I tried to open it but the door was locked. The pathetic meowing continued, and I decided to try and break the door open. Giving the door a hefty kick, the flimsy lock broke and there was a small, emaciated and very frightened cat. It must have been surplus to the needs of whoever owned the house, and had been locked in here to die. It was ravenously hungry, and took great chunks out of the cheese I offered it before heading very unsteadily off across the plains, trying its best to run, but managing little more than a pathetic trot. It must have been terribly dehydrated, and I rather doubt it survived. It occurred to me to keep it as a pet. It would have provided welcome company, but somehow I couldn't see the arrangement working.

★

Just below the ruins, I chanced upon a scattered colony of the very rare *Asteriscus schultzii*, growing from the bare hillside. This low growing, daisy-like plant has pale yellow flowers about 2.5cm across with a bright yellow centre. In Fuerteventura it grows only in the area north of a line between La Oliva and Cotillo, but can also be found in some parts of Lanzarote. Until recently it was considered to be an endemic to these two islands, but has now also been found in Morocco.

An hour beyond Gavias del Carce was a semi-circular wall on the hillside: a shooting butt for the autumn Barbary Partridge hunt (Box 9). I hadn't seen any bustards, despite repeatedly stopping to scan the arid terrain. I decided to rest here a while and recover my energy as I was already beginning to flag. I was feeling lonely and daunted.

★

Giving up on the bustards, I carried on towards La Oliva. In front of me was Rosa de los Negrines with its fields, fig trees and scattered palm trees: a veritable oasis in this stark landscape and one of the finest spots on the island. It lies in a slight hollow at the edge of a lava field, in a place where water naturally collects after the rains. Behind is the grey-green crater of Montaña Arena that dominates the western end of La Oliva (cover photo). To complete the picture, there is an old *taro*, a small circular tower looking rather like the stump of a windmill, but actually an old agricultural storage facility. These structures are found all over the island, though most are not as well preserved as this one.

The eastern end of the farm was largely unused in the early 1980s, and I often had lunch there in the shade of the trees. After rains the cultivation was bright green, splashed with the showy yellow and white flowers of the Crown Daisy *Chrysanthemum coronarium*, a common Mediterranean species that flourishes in fields and along the roadsides on the island. There was a cluster of abandoned but still intact buildings and even a little chapel with a tiny bell tower that was the favourite perch of a Southern Grey Shrike, a showy grey, black and white bird that is common on the island (Plate 38). At times there were also migrant warblers in the trees, swallows flitting over the fields and perhaps a gaudy flock of Yellow Wagtails – all attracted by the relatively lush vegetation.

In March of the first year of my studies, two birdwatching friends came to stay for a couple of weeks. Dave Shirt and I were at college together, and he had also been a member of the 1979 expedition. Howard Taffs was a birdwatching friend from school that I have been in touch with ever since. They were lucky enough to arrive on the island just as large numbers of migrant birds appeared. The fields and bushes at Rosa do los Negrines were alive with birds, including two species that had never been seen before in the Canary Islands, a Bonelli's Warbler and a splendid male Cretzschmar's Bunting. Although the Bonelli's Warbler is now known to be a scarce but regular passage migrant in Fuerteventura, the bunting remains the only one ever to have been seen in the islands. It is a handsome bird about the size of a sparrow, but rust-coloured with a striking blue-grey head.

The area has now been brought back into active use, and as I walked past there were noisy dogs tethered at the entrance. I gave the fields a wide berth and re-joined the mule track at the point where there is a dense growth of the succulent spurge *Euphorbia regis-jubae* in a slight gully. It was pleasing to be amongst bright green vegetation after spending the morning out on the barren hills. I could hear the seed pods of the Euphorbia popping in the hot sun, and lizards scuttled for cover as I pushed my way through (Box 10).

Beyond the Euphorbia clump, the old mule track climbs steadily towards La Oliva. I stopped for one emotional last look back over my old bustard study site. Then a tourist spoilt my reverie by driving noisily past on a motorbike, before turning round and covering me in dust as he sped back towards the road.

Box 9 Hunting and the EU Birds Directive

The Barbary Partridge was apparently introduced to the island from North Africa in the early years of the twentieth century as a game species. Traditionally, the opening day of the hunting season in September was probably as important in the calendar of Majoreros (natives of Fuerteventura) as the glorious 12th of August is to the grouse-shooting fraternity of Scotland and the uplands of northern England. However, the partridge has been overhunted and is now rather scarce. Rabbits are also hunted, although these are none too common on the island either. One can't help wondering if the occasional bustard also gets shot from time to time: it must be a very tempting target. They used to be highly prized for their meat, not surprisingly given the all too frequent recurrence of severe droughts and famine, but they are now fully protected by national and European legislation.

The European Birds Directive has had an important impact on bird populations throughout Europe, and has slowed the decline of some of the continent's most threatened birds. Amongst other things it requires member states to establish Special Protection Areas (SPAs) for important populations of threatened birds. In Fuerteventura there are no less than five such areas, one of which includes my old bustard study area. Of course, one of the principle species that the area has been

---

designated for is the Houbara. It is gratifying to know that my studies helped to establish the area as a protected zone.

Box 10 Lizards

The common lizard on the island is *Galliotia atlantica*. It is the smallest of the endemic Canary Island lizards growing to less than 10cm in length, and superficially not unlike the common lizard found in Northern Europe. It occurs only on Gran Canaria, Lanzarote and Fuerteventura. It is abundant on Fuerteventura, where it appears to form the main prey of the Kestrel. On the whole it is a plain little lizard although the mature male has blue-green stripes on either side of its body. There are no snakes here but there are three other reptiles. A much bigger lizard is the Giant Gran Canary Lizard *Galliota stehlini*, which can grow up to 50cm long. It is endemic to Gran Canaria, but turns up on Fuerteventura from time to time as an escapee from the inter-island cargo ships, and has perhaps also been deliberately introduced in the past. In all my wanderings around the island I have never seen it myself.

The other two species are native. The little Fuerteventuran Gecko *Tarentola angustimentalis* is very like the common Mediterranean species, but is found only on Lanzarote and Fuerteventura plus the adjacent islets, including Lobos. Like other Geckos it is common around buildings, and can often be found near external lights around the villages and tourist complexes. Finally, and most interesting of all, is the endemic skink *Chalcides simonyi* which is also found only on Lanzarote and Fuerteventura, although a similar animal is found on the adjacent coast of Africa. It leads an almost entirely subterranean life, living primarily under rocks and is rarely seen. It is a handsome shiny bronze colour with the short, stumpy legs that typify skinks. I have never seen one in the wild, but one was brought to the Biological Station when I was living there.

NINE

La Oliva

The road into La Oliva from the west skirts the edge of the *malpais*. There are houses built on the lava slopes north of the road, whilst the land on the other side is flat and partly cultivated. Back in the 1980s some old stone houses had already been abandoned, but others were still inhabited, often by old crones in black who wore wide-brimmed straw hats. Even the inhabited houses looked to be in poor repair, with earth yards and perhaps a few pots of geraniums. Today, nearly all the old houses are wasting away and there is a rash of new build, including some rather fancifully designed houses that testify to new wealth, no doubt derived one way or another from the tourist boom. New also are decorative gardens: one was full of Bougainvillea, tall cacti and other succulents and another had a fine olive tree, Mimosa and even a Dragon Tree. Beautifully landscaped gardens are another sign of affluence. Times have very definitely changed.

The last few hundred metres before the centre of the village were barely recognisable to me. The jumble of old buildings, many of which were already redundant in the early 1980s, has been swept away, and new apartment blocks were being built. Opposite these is a heavily restored tithe barn, the only one in good condition on the island, originally built in 1819. It was restored to house the La Cilla museum, which describes the old agricultural practices on the island. This is one of a range of new museums, set up by the island council, that have appeared around the island over the last decade or so to give tourists somewhere to go. Not that I am against them: we need to remember how things were, and the museums are invariably well presented. The text accompanying the displays is primarily in Spanish, which is perhaps surprising given that the first language of many of the visitors will be English, German or French, but most of the museums have leaflets with information in different languages.

The other agricultural museum on the island is the Ecomuseo de la Alcogida at Tefía, which, despite its name, is not a museum of ecology. It is essentially a collection of buildings restored to illustrate rural life and crafts in the early part of the twentieth century. It is well done and also worth visiting.

★

On the main street I bought some provisions in a small shop and arrived gratefully at the Bar Hijos de Suarez. This is primarily a bar for locals, and is built more or less where my old apartment used to be, directly opposite the church. The Suarez in question is my old landlord Don Pepe. In the previous year, when I came here with my wife, the young man who served us bore more than a passing resemblance to Pepe, and was no doubt one of the sons (*hijos*) in question. By some bizarre coincidence, the sugar sachets that came with the coffee had pictures of Houbara Bustards printed on them. The Spanish Ornithological Society had sponsored a series of sugar sachets! I now have an old Turkish delight box with five Houbara sachets, plus one each of Fuerteventura Chat, Trumpeter Finch and a Laurel Pigeon (from the western islands).

It was two days since I had had a proper meal, so I ordered two large pork chops with chips and a large salad. I collapsed at a table with a bottle of water and waited for the food to arrive, grateful for the cool darkness of the bar. Having sat down, I realised that I was not only feeling really tired but that my head was spinning slightly and I felt faintly nauseous. I knew these symptoms were likely to be the result of dehydration. I clearly hadn't been drinking enough given the hot weather and the physical effort I had been making. I needed to drink lots of water and rest as long as possible before continuing.

If it was dehydration, that would resolve the problem. If not, then I would have to revise my plans, but I didn't want to think about that at the moment so I pushed the thought to the back of my mind as much as I could. I was, nevertheless, worried. I popped back to the shop to buy another litre and a half of water. Although I had no idea how I was going to carry it, the next shop was a day's walk away and I couldn't risk running out of water. With a good deal of effort, I just managed to find room for everything in the rucksack, although it was a very tight squeeze indeed. I really should have purchased a larger one.

Whilst waiting for my lunch, I studied the photographs on the walls. They had been taken in the early part of the twentieth century, and showed towering hay ricks and people ploughing with camels and asses. The same pictures are to be seen in many of the bars on the island. When I first visited the island in the early 1980s this way of life was disappearing fast, but I did occasionally see a camel pulling a plough, and many of the houses had the characteristic domed haystacks. Goats were everywhere then, and were still the mainstay of the economy in La Oliva and all the inland villages.

★

Although I was not feeling too good, once my meal arrived I ate as much of it as I could and drank several bottles of mineral water, then wandered out. The village square, to one side of which sits La Oliva's large and distinctive church (Box 11), is now a rather smart public space. In the past it was a pleasant enough informal space with a few palms and other trees, but it has recently been re-landscaped to match modern tastes, and enlarged to include what used to be part of the road. The whole venture cost more than half a million Euros, a lavishness that would have been unthinkable thirty years ago. John Mercer, in his musings about the future of the island, states that the entire budget for the island council in 1960 was just £11,000, or about 12,000 Euros at 2009 rates. The island councils are so much better off than they used to be, thanks to the tourist boom, and no doubt money from the European Union also comes in handy. If money means anything, this would appear to be Fuerteventura's golden age.

★

On the south-east margin of the town is the handsome Casa de los Coroneles, which is perhaps the most imposing building on the island (Box 12). The fields to the south of it are some of the lushest on the island. They lie in a fertile basin that receives relatively large amounts of run-off water from the mountains to the south. All this greenery attracts birds, and I always make a point of visiting the spot when I am on the island. Amongst other things, it is particularly good for Hoopoes, fanciful pink, black and white birds with a floppy, almost butterfly-like flight. On one memorable occasion, when I was on holiday with my daughters Clare and Katy, there were more Hoopoes here than I have ever seen in one place. I have a wonderful photograph with no less than fifteen of them in various poses.

Having completed my meal, I ambled down to the fields, but there were few birds to be seen, just the usual Lesser Short-toed Larks and Berthelot's Pipits amongst the crops, and a single Corn Bunting giving its jingling song from the top of the fence.

I followed a track towards a line of stately palm trees. In the days when the economy of La Oliva was heavily dependent on goats, the flocks heading up this track at the end of the day were a sight to behold. I can still hear the bells of the lead animals clonking as they made their way home, desperately trying to snatch a last mouthful of weeds as they were driven on. There are three methods employed to hasten the goats along at such times: vocal threats, the ever-present dog, and for wayward animals a well-aimed stone.

Near the palm trees, tiny African Grass Blue butterflies flitted over the sparse vegetation. These diminutive, dull blue butterflies are recent colonists from African. Unknown here until the 1990s, they are now found in the more lavishly landscaped tourist complexes as well as around larger cultivated areas (see Box 13 for information on the common butterflies of the island).

There is a ruined farmhouse by the palm trees. It was already derelict thirty years ago, and I have a picture of it in my study. In those days the roof was partly intact and the wooden door and shutters were still there. Today only a few bare roof beams survive, and the windows and doors are little more than gaping holes. Beside the ruined house there is an *aljibe*, a deep rectangular cistern for storing water, complete with roughly fashioned drinking trough. In the relatively wet spring of 1980 it was full of water, and in the late evening there was a deafening chorus of Stripeless Tree Frogs, that I could hear a kilometre away from the roof of my home in La Oliva. These little green frogs are about 5cm in length and they hide in vegetation. They hide so well in fact, that I never did manage to see one. The other amphibian on the island is the larger Marsh Frog, which inhabits the few reservoirs where there is near-permanent water.

That same wet spring there were several fields here with a really lush growth of grass and flowers: it was liberally dotted with crown-daisies. This provided a nesting habitat for two or three pairs of Quail. The distinctive 'whet-my-lips' songs of these secretive little game birds could be heard day and night for a few weeks. Quail are normally hard to see because they habitually stay in cover at all times, but these birds occasionally called from the tops of the earth banks that were in full view over the top of the grass.

The status of the Quail in Fuerteventura is something of a mystery. They are apparent once every few years, when there has been enough rain to produce a really good flush of vegetation. Then they seem to disappear again. Some have suggested that they are partial migrants, and leave the island when the rains fail. My own view, having flushed birds from odd places around the island over the years, is that they simply disperse to wherever they can find food outside the breeding periods, but are rarely observed because of their secretive habits.

The palms themselves have always had a pair of Kestrels nesting in them and I was pleased to see that they were still there. The palms also make a good home for a noisy colony of Spanish Sparrows. This handsome sparrow, the male with its dapper black waistcoat, is the only species on the island. It is common around habitations everywhere on the island, but is also found in quite remote areas where there are suitable nest sites. It seems happy enough to share its home with the Kestrels.

★

I found a stone platform in the ruined house to sit on. I was now really apprehensive about the next stage of the walk. For the rest of the day and much of the next I would be alone in the mountains, sometimes in places I had not been to before. What I was about to do was one of the most challenging things I had ever done and I had to admit that I was genuinely scared. I sat quietly and forced myself to meditate for a while. It was difficult to still my mind, but this did calm me somewhat. I checked my mobile and found that my wife had sent me a text the previous night saying simply 'night, night'. That was remarkably comforting. It seemed sensible to wait until the day had begun to cool before continuing my walk, so I read the wheelbarrow book for a while, although retrieving it from the bulging rucksack was a logistic nightmare requiring remarkable persistence! I tried to read it for about fifteen minutes, but my head was full of the difficult journey ahead.

The first task was to climb the very steep slope onto the ridge to my south. I had never got round to doing that when I lived here, and now I had to do it with a heavy pack and with sore feet. I planned to drop down into the deep Vallebrón valley beyond, and then up onto the ridge that leads to Muda, the highest mountain in the north of the island at 689m. Did I have enough provisions to keep me going through the physical challenge? Where would it be possible to pitch a tent in these bare rocky hills? This was it, the easy stuff was over. Having come this far, turning back at the first real hurdle was not an option. I had to give it a go, even if I failed. I picked up my pack and headed for the ridge.

Box 11 La Oliva church

The church was built immediately after 1708, when the military governor of the island was established in La Oliva. Perhaps this explains its fortress-like appearance, and it is certainly an imposing building (Plate 26).

The most conspicuous feature of the church is its muscular tower, which has four levels of rough faced stone. This contrasts markedly with the rest of the building, which is whitewashed. Much of the lowest stage, and the corners of the upper stages, is made from well-dressed rectangular blocks of lava, but the spaces between them are in-filled with small, uncut stones of various colours. Except for the belfry at the top, there are no openings in the fortress-like tower except for a single rectangular opening high on the west side. The belfry has two tall, round-topped openings on each face. An outer staircase on the south side leads to a door which gives access to the four bells. There are two large bells, one in each of the south facing openings, and two smaller bells in the west facing openings. In common with church towers the world over, feral pigeons have taken up residence in the belfry. The general appearance of the tower is not unlike that of the sixteenth-century church of El Salvador in Santa Cruz de la Palma, especially in the shape of the bell openings, so perhaps it is modelled on that building, although it is certainly more roughly built.

The other main feature of the church is its three aisles: it is an unusually large church by Fuerteventuran standards. Each of the aisles has its own pantiled roof. The main entrance is in the central aisle, which has a huge triangular-topped doorway, with a massive, round-topped wooden door. As is typical of the churches on the island, there are few windows, and these are set high in the walls. The church is dedicated to Our Lady of Candelaria, the image of which is paraded through the streets on 2nd February.

Box 12 Casa de los Coronelles

From 1708 until the early years of the nineteenth century, La Oliva was the military capital of the island, and was also to some extent the administrative capital. At that time, La Oliva must have been a distinctly more important place than it is today.

The establishment of the military governor in La Oliva required the construction of a suitably grand building: the Casa de los Coronelles. This stands a respectful distance away from the centre of the village and its church, overlooking the fertile plains to the south-east. It is a fine building, and one of the most important historic buildings on the island.

It has two storeys on a square plan with a central courtyard. The façade facing the village is impressive, with a fine pan-tiled roof and castellated towers at either end as befits a military residence. An important feature is the windows of the upper floor, which have finely carved wooden balconies. Inside, there is a splendid wooden gallery around the courtyard, supported by elegant wooden pillars set on carved lava blocks to keep the wood dry at the base. The gallery is accessed from the courtyard by a flight of steps with a rather fine trifoliate archway.

There is some confusion in the guidebooks as to when the

house was built. Some say it is seventeenth century, others that it is eighteenth century. It looks to me as though there may be some truth in both claims. A house was first built on the site by the Cabrera Bethencourt family in the second half of the seventeenth century, and it is apparently their coat of arms that is carved into the stone over the main doorway. However, when the military governor took up residence in 1708, the building was substantially rebuilt and enlarged. We can guess that the castellated towers were added at that stage. By the time the last governorship ended in 1870, the governor was spending most of his time away from the island, and perhaps the condition of the house had already deteriorated by then.

In the early 1980s, the house was empty and in a state of serious dereliction, having been little used for most of the previous hundred years. At that time the façade was covered in crumbling red and orange paint (the colours of the Spanish flag), and the lower rooms were home to large numbers of goats. However, the State purchased the house in 1994 and made possible the restoration that has now taken place.

The change in fortune of this fine building is another example of Fuerteventura's golden age. On 24th November 2006, no lesser persons than King Juan Carlos and Queen Sofia of Spain came to La Oliva for the official opening of the splendidly renovated building during a visit to the Canary Islands. In spring 2007, when my wife and I visited it, it was in pristine condition with a fine new coat of cream wash on the outside. Judging from my old photographs, the restoration has been substantially true to the original features. Of course it is one thing to restore a building but quite a different problem to give it a use, without which there is the risk of it falling into disrepair again. Only time will tell what will become of the place.

There are two other significant buildings associated with the Casa de los Coronelles that receive rather less acclaim in the guide books. A rather dilapidated little building to the south is the remains of the dovecot that presumably ensured a supply of fresh meat even in the worst of times. Inside there are dozens of square pigeon holes set into its walls. There are no pigeons now; their descendants evidently prefer the safety of the church tower! The second is a low, whitewashed building to the right of the road as you approach Casa de los Coronelles from the centre of town. This is known as the Casa de Capellán, or the chaplain's house. At first glance it looks to be of no great interest, but the stones around the windows and door have fanciful carvings that are reminiscent of the extraordinary carvings on the church at Pájara. The carving over the doorway includes a cross and two crested birds' heads, which might either be cockerels or perhaps peacocks. According to guidebooks, it is around 200 years old, in which case its decoration is derived from the church at Pájara but it is not contemporary with it.

Box 13 Common butterflies

There are not many species of butterfly on the island: islands are generally poor for butterflies anyway, and Fuerteventura is too dry and inhospitable for most. The two commonest ones are African desert fringe species: the Green-striped White and Greenish Blacktip. After the rains have come, whilst the annual plants are in flower, these are everywhere, from the shore to the tops of the highest and most barren peaks.

As its name suggests, the Green-striped White is a white butterfly, and the green stripes that give it its name are on the underside of the hind wing. The similar Bath White is also found on the island, but has green blotching on the under wing rather than stripes. The Greenish Blacktip is less well named; it is a pale yellow butterfly about the same size as the Green-striped White.

The only other butterflies that are common on the island are Painted Lady, Red Admiral and Clouded Yellow.

Ten

Onwards and upwards

As I set out for the hills, there were swifts screaming high above. These were Pallid Swifts, which nest all over the island. They are a little larger and paler than the Common Swift, and their calls are slightly different: almost two-tone and finishing on a lower pitch. They reminded me of my father, who died more than ten years ago. On my way home from school one day the swifts must have been particularly noisy, and I commented that I loved all bird songs, but not the screaming of swifts. "I do," said my father, "it is such a wild, free sound." He was right: there is nothing as elementally free as a party of screaming swifts flying wildly through the sky.

In some ways this walk was also an expression of freedom. I had a tent and provisions: I was not dependent on anyone else for the next twenty-four hours, and I would have no contact with people at all until I arrived at La Matilla some time the next day. And yet in another sense, I had no freedom at all: I had set myself this task and had to go where I feared to go. There was no choice now but to head into the unknown. I was stirred from these thoughts by a loud 'coo-lee' call coming from the big fig I was passing just then. A handsome Southern Grey Shrike was perched only a couple metres from me, watching from behind his black bandit's mask. Somehow I found it very reassuring.

★

First I had to get to the top of the ridge that dominates the southern skyline from La Oliva. There was no path to follow, and it looked formidably steep. It was hot work, but the ascent was going reasonably well until fierce dogs started to bark and snarl behind me. I put on an added spurt but soon realised that they were in fact all tethered. Unfortunately, the extra burst of effort caused a blister that had now developed on the sole of right foot to burst, causing significant discomfort.

A bigger problem than my feet, though, was the wind. The ridges that run east-west create a barrier to the prevailing wind and often have violent winds towards their summits. Halfway up, the wind was already very strong, the fierce gusts blowing me around on the unstable slope. There was a fine view back over La Oliva, but my mind was focussed on how to get up the last 150m. It had looked steep from below, but the slope in front of me now seemed almost impossible. Apart from being incredibly steep, it was covered in loose rock and with the unbalancing effect of the fierce wind it was a daunting prospect. I edged tentatively forward, but it was even worse than I had feared. Several times I lost my footing and the heavy pack nearly pulled me backwards, the wind grabbing at me and threatening to destabilise what tenuous grip I had. I now realised the folly of what I was trying to do: I was at risk of serious injury. The only way to make upward progress without losing my balance was to crouch forward like a Neanderthal. Thus hunched, I took one step at a time, testing the rocks and concentrating on maintaining my balance. I found myself adopting the mantra 'slowly, steadily onwards', which I repeated at frequent intervals to help steady my nerves.

After what seemed like an eternity, the slope slowly began to ease. Having gained the ridge it was a short walk to the summit of Morro Carnero (508m). I was greatly relieved to have made it in one piece, but of course I now had to find a way down. The intention had been to maintain a southerly heading from here straight across the Vallebrón Valley. But the valley before me was virtually a chasm, the slope even worse than the one I had just come up, almost vertical in fact. There was no way I could get safely down it with the pack. In fact the whole idea of crossing the mountains to La Matilla seemed utter folly. It would be far too dangerous for me to undertake on my own. There was nobody in these hills and if I fell or broke an ankle I would be in serious trouble. I decided there and then that I must find my way down to the coast and carry on south through Puerto del Rosario. The walk would be much less interesting, dull in fact, but at least I would be safe.

I headed down the much gentler eastern slope towards the village of Caldereta. It was still quite steep but not dangerous, although my feet were horribly sore and I was tired after the long walk and the fear that I had just experienced. The land below me looked quite flat and through the binoculars I spotted a patch of greenery with a few fig trees that looked far enough from roads and buildings to allow me to camp unnoticed. On my way across the valley floor I flushed a Quail from a patch of grass, and there were Corn Buntings singing. I began to recover my composure.

The place I had chosen proved to be an ideal sheltered spot, and in stark contrast to last night's wind-swept camp in the sandy wastes south of Lajares. It was evening by the time I had the tent up and everything was inside. I was out of the wind and enjoying the peace and warmth. There were crickets chirping soothingly close by in the long grass and a Berthelot's Pipit singing. I had found the perfect spot and now I could relax and take stock.

First things first, I took off my boots and socks and inspected my feet. They were both seriously blistered now and I wasn't sure how much more walking they could cope with. I had to decide what to do from here, so I studied the options on the map. If it was too dangerous to cross the mountains, the only alternative was to walk down the east coast. I would have to walk as far as Puerto del Rosario, but from there I could get a taxi to my hotel in Fustes. The road from Puerto to Fustes is very busy, and the heavy traffic would make it too unpleasant to contemplate walking along. Sadly, though, this would be the end of my dream of walking the length of the island. I had a big decision to make.

In the fading light of early dusk there was an anxious moment when a man came in a truck to collect herbage for his livestock. Would he come to the patch of greenery where I was camping and if so would he spot my tent tucked under the wall? What would his reaction be? But eventually he drove off and it was soon completely dark. I could finally relax in the snug warmth of my sleeping bag. All the worries of the day were behind me and the soothing sound of the crickets seemed to work some magic on my frayed nerves. A Stone-curlew called nearby. At first there was a bit of traffic noise from the road 500m away, but it died down as the evening wore on. In a rather better state of mind, I reflected on the day's events and wrestled with the decision about what to do the following morning. It felt as though I had learnt an important lesson: one that many must have learnt. Choose your dreams carefully, because if you decide to follow them through you will have to live with the consequences! Dreams can lead you astray and turn into nightmares. It is also important to know your limits and make decisions accordingly. I reminded myself that the essential thing was to enjoy the trip. Only so much discomfort and risk would be compatible with that objective.

At some stage the crickets must have lulled me to sleep, but I woke at two a.m. and lay there for two hours thinking about what to do. Then my decision was made, I would walk to Puerto, get a taxi to Fustes and hire a car for two weeks. There was no point going on if I couldn't get up and down the mountains, and my feet were already a mess. What would they be like after two more days of hard walking on rocky ground? Once my feet had recovered, I could drive to wherever I want and do all kinds of long walks with a much lighter pack. This was a big decision, but I couldn't see any other way forward. I was glad that I had at least tried to walk the island, and it had already been a real experience, but I was determined to enjoy being here. In the morning I would turn my back on the dangerous mountain route, and take the easiest route to Puerto. Having made my decision, I went back to sleep.

At dawn the following morning I broke camp at first light, stuffed everything back into the rucksack, and headed for the mountains! There was, I decided, nothing to lose in trying to get over the mountains to La Matilla. If I couldn't make it up the slopes I would turn back, and at least I would have tried.

★

So I doubled back the way I had come the day before, heading for the Vallebrón valley. It took me twenty minutes to regain the ridge, from where I had a fine view up the valley to the scatter of whitewashed houses that is the village of Vallebrón. On the other side of the valley was the ridge that I needed to climb, with the peak of Muda (689m) at its western end. My intention was to climb the ridge at its lower, eastern end, at the point known as Morro de la Pila, then walk along the ridge to the peak of Muda itself. Studying the slope carefully, I thought I could see a reasonable way up, but first I had to find my way down into the valley bottom.

The slope was entirely composed of horrible loose rubble with no trace of a path of any kind, but it proved to be just manageable as long as I gave the task my complete concentration. Part of the way down, a bird flew up from rocks at my feet. It was a Fuerteventura Chat rising from her nest under a rock. These deep eastern valleys provide good habitat for chats. There were more chats further down the slope, and overhead a noisy pair of Ravens on their early morning patrol of the valley.

I reached the bottom of the valley at a point where there were seven fig trees. There was a rough vehicle track, but all was quiet apart from the birds. A noisy Shrike didn't approve of my intrusion into his patch of fig trees, a Barbary Partridge called somewhere on the hill above, and a Corn Bunting was singing. This was the peaceful Fuerteventura that I knew and loved and I was in good spirits.

Beyond the track, and after successfully negotiating a series of shallow gullies and walls, I started up the steep incline leading to the ridge. I could see the village of Caldereta away to my left, with the sea beyond. I was beginning to enjoy myself, and so far at least I was very glad to be making the attempt and not plodding down to the coast road in failure. Using the hunched technique I had learnt on the dangerous slope the day before I made slow but gradual progress and found my way onto the ridge without much difficulty. Away to the south I could now see the sprawling mess of Puerto del Rosario in the distance, with the airport beyond it, and in the far distance that I could just see Fustes, where I hoped I would be resting in a cosy hotel by the end of the following day.

I could see the peak of Muda five kilometres away at the far end of the ridge (Plate 6). It looked terribly steep and daunting. One annoying characteristic of ridge walks is that they always look so much easier from down below than they really are. This one was no exception, and I could now see that there were several substantial hills between me and the peak. I had, however, done this walk before, though admittedly not with a heavy pack.

Just after nine o'clock I found a cairn to shelter behind and sent a text to say that all was well. There was a message from my wife about getting her hair done: life was going on as normal back home in the real world. I finished the last of the big bar of whole nut chocolate she had given me. Two Lesser Short-toed Larks were singing over the high plateaux. There were lichens on some of the larger rocks, but otherwise virtually no vegetation at all apart from a few stunted annual flowers a centimetre or so high, and it was difficult to understand what these birds were living on. One of the larks was a particularly good mimic, treating me to snatches of Linnet, Berthelot's Pipit and Pallid Swift songs.

A bit further along I came to a ruined house. Quite why anyone would want to build a house up here is hard to imagine. As far as I could see there was no track linking it to the village of Vallebrón in the valley below, but perhaps the ridge had sufficient vegetation to sustain a small flock of goats. At home I have some bits of purple glass and coarse pottery found here during a previous visit, and there were more bits of the same still scattered on the ground. It seems likely that the house was ransacked after it had been abandoned. Next to the house was a small enclosure where livestock would have been kept at night. There were nettles growing from the base of the walls and even a Henbane *Hyoscyamus alba*, indicators of the high fertility that comes from decades of use by livestock. Just below on the hillside, a beautifully constructed near circular stone cistern had been built, and much of the clay lining was still intact. It was evidently designed to gather runoff from the hillside.

★

The very top of Muda had now disappeared into the clouds, as had the tip of Aceitunal, Muda's sister mountain south of La Matilla. To my left was the deep valley of *El Sabio*, the wise one. I wondered who it was that this uninhabited valley was named after. It is one of the remotest places in the northern part of the island, so perhaps a hermit lived here at some time, though there appear to be no caves. The steep slope leading down into the valley has a fine stand of *Euphorbia balsamifera*. This plant is bigger than its commoner relative *E. regis-jubae*, with the same succulent stems and branches and rosettes of darker green leaves. It is typically two metres high and several metres across, so walking amongst a colony of them is a little like walking through a strange diminutive forest, a very welcome feeling after so much bare aridity. On Fuerteventura it forms the natural vegetation on dry south-facing slopes such as this, where the rainfall is typically less than 150mm. Sadly though, it has been lost from large areas due to human activity, and in most places it has been replaced by scattered growths of lower growing shrubs.

To my right the houses of Vallebrón were scattered over the valley floor 300m below. Hemmed in on three sides by steep slopes, it is the closest there is to a mountain village in the north of the island. Although still quite isolated, there are grand new properties springing up even here.

★

The narrow ridge approaching Muda is a wonderfully exhilarating walk, with steep drops on both sides and the fearsomely steep looking Muda itself rearing up in front. The last two kilometres includes a series of small peaks over 550m. Intriguingly, one of these (disappointingly, an unimpressive rounded one) is called Pico de Don

David, although whether the David in question was as British as his name suggests I am unable to say.

Up here, the annual rainfall is above 200mm per year, about twice the amount over much of the island. As hard as it is to imagine now, the natural vegetation above 500m is dry evergreen woodland dominated by the endemic olive tree *Olea cerasiformis*. The trees have long since gone; they were too useful to mankind. They must have been especially valuable to the pre-Conquest islanders who had no way of importing wood from elsewhere as they had no ships. In addition, constant grazing by goats through the centuries has prevented regeneration, and today the high ground is as bare as the rest of the island, or very nearly so. But on Morros Altos, which is the last peak before Muda, there is a lone surviving Olive which sprouts from the rock and grows not upwards but due south, all upright growth having been nibbled by the goats. It could hardly be described as a tree: it is just one small branch a few metres long. But it may well be very old, a bonsai formed naturally by the fierce wind and the nibbling of goats, the last remnant of an ancient forest. I was pleased to see that it not only had healthy leaves but also a few flower buds.

There are quite a few other plants up here that are not found further down. One of the more obvious ones is the Milk Thistle *Silybum marianum*, with its big, angular leaves patterned with white, and the small, yellow-flowered mignonette up here is the eastern islands' endemic *Reseda lancerotae,* a common plant on the high ground throughout the island.

★

At last there was nothing between me and the peak of Muda, rising very steeply for almost 200m, which made yesterday's terrifying climb look like a vicar's tea party. The sun was now threatening to break through, so I took of my fleece, smeared sun cream on my exposed hands and face and put my hat on. I was ready for the climb. I would just take it gently, repeating my mantra to keep my focus. Just as I was about to start I noticed a perfect shell at my feet. I picked it up for good luck, and I have it in my hand now as I write. It is a cone shell 35mm long and 20mm wide at the wide end, which tapers to a blunt spiralled point. There is a long, slit-like aperture where the animal would once have extended its foot. It is smooth and white with age. How did it get to be 550m above sea level? There would simply be no point in any bird carrying it up here, so it must be that the sea bed this shell was resting on was thrust upwards during some period of intense volcanic activity several million years ago when these mountains were being formed.

I began the steep ascent, hauling myself up the slope. At my feet, scattered between the jumble of loose rocks, the ground was now dotted with a variety of yellow, mauve and purple flowers. Amongst them there was quite a lot of Yellow Pheasants-eye *Adonis microcarpa,* an attractive member of the buttercup family that is apparently widely distributed in the mountains. As its name suggests, the petals are yellow, but below these there are red sepals, and the centre of the flower is black. The whole effect is very attractive.

Slowly, as I gained height, a spectacular view of the desolate country below unfolded to either side. To the north, the volcanoes I had walked amongst two days before came into view in the distance, and in the far north the sand dunes of Corralejo and the island of Lobos beyond. Lanzarote, though, was lost in the haze. To the south, the mountains of the central massif around Betancuria came into view for the first time. I would not reach them for several days yet. I had nearly reached the top now, but the last bit was very steep. Fortunately, close to the top there was an exposed slab of basalt largely free of scree, and by scrambling up this rock I managed to avoid the worst of the loose material.

Finally, just before noon, I reached the huts by the summit, behind which I was grateful to shelter from the fierce swirling wind on what was the only level ground on the whole mountain. At the very top there is a mast supported by a number of sturdy cables, and the wind blowing through these made a fearful roaring sound: this is not a restful place. After a short while I edged round to the south side of the mountain, where there is a vertical drop. I took in the awesome view for a few minutes, gazing out towards the Betancuria mountains.

Then I started my descent, picking my way across hazardous steep scree to the vehicle track that allows access to the mast from the south. This track would take me down to La Matilla, and hopefully a decent lunch.

Coming down over the scree was difficult, and on one occasion I did stumble and fall. After that I tried the technique we were taught in Lapland for going safely down an icy slope on skis, turning sideways and descending crab-wise. Despite the lack of skis this worked very well, and I would recommend it.

Just after reaching the safety of the track, two adult Egyptian Vultures came into view below, boldly patterned black and whitish birds, with a diamond-shaped tail and bare yellow skin on the head. They came quite close as they followed the air currents up the slope towards the top of Muda, a wonderful sight.

Beside the track, there were lots of the heart-shaped leaves of the Friar's Cowl *Arisarum vulgare* growing from a rock outcrop. This pitcher plant is not uncommon on rocky places in the mountains, although I have yet to see one with the curious hooded flowers that give the plant its name.

For some reason, going down was more painful on my feet than going up, but the call of lunch was great, and I found that the best way to get down this steep track without slipping on the loose grit surface was to adopt a strange loping trot. Thus propelled, I made good progress and before very long emerged on the road in La Matilla. There was a short, steep bit of road that taxed my aching legs and then, just an hour after leaving the summit, I found myself at the Bar Restaurant La Matilla. To my delight it was both open and happy to serve lunch. I ordered goat with chips and salad, took the boots off my sore feet and slumped triumphantly into a chair with a bottle of cold water. There was a wonderful view of Muda from the window (Plate 28).

Eleven

Walking to Fustes

Over a well-earned lunch in La Matilla, I considered the options for the next stage of my journey. La Matilla sits between the two highest peaks in the north of the island, Muda and Aceitunal. From here, my original intention had been to ascend Aceitunal, then walk along the four-kilometre mountain ridge to the south, dropping to the village of Casillas del Angel to camp somewhere south of there. I now abandoned this plan and took a more direct route to the coast at Fustes. There were two reasons for this decision. Firstly, although I had now proved to myself that I could walk in the mountains with my heavy pack, at the current rate it didn't look as though I could cover the remaining thirty kilometres to Fustes by the following evening. Secondly, now that I was close to Aceitunal, it looked even steeper·than Muda, and I wasn't sure I could safely get up it and on to the ridge beyond. My heart didn't warm to the idea of walking on roads, but my feet were already in a mess, and I was keen to get to Fustes as quickly as possible so that I could rest them.

The decision made, I paid for my meal and wandered over to the cool shade of the bus shelter across the road. I was relaxed after successfully completing the testing walk through the mountains; I had eaten well and was confident now that I could complete the walk. I took out the wheelbarrow book and read on and off to pass the time. An African Blue Tit sang from a dead agave flower just across the road. Until recently, this bird was considered to be a race of the

European Blue Tit. Now, however, the birds in the Canary Islands and North Africa have been recognised as a distinct species. Unlike European birds they have bold black and white head patterns, are bright yellow below and are darker, slate-blue above. Their song is much louder and more strident. It is not particularly common in Fuerteventura, and is found mainly in the mountains and around villages and towns where there are enough trees and bushes for it to feed in. It also occurs in extensive areas of *Euphorbia* scrub, even in the highest parts of the island.

After an hour or so I carried on towards Tetir. Growing beside the road, I came upon three spikes of the mauve-flowered Branched Broomrape *Orobanche ramosa*. Broomrapes are unusual plants in that they have no chlorophyll and therefore cannot photosynthesis their own food as do virtually all other flowering plants. Instead they are parasitic on other plants. Some broomrapes live only on a specific species, but this one seems to parasitize a wide range of plants. The ones here seemed to be growing in association with a Launaea shrub.

Around mid-afternoon I arrived footsore in the pretty town square in Tetir. I found a bench where I took the pack from my aching shoulders and gratefully removed my boots. Then I sketched the lovely church of Santo Domingo de Guzman. It has a fine tower, perhaps the best on the island, somewhat suggestive of a five-tiered wedding cake, with each of the gleaming white stages a little smaller than the one below. The tower is charmingly feminine, and in complete contrast to the masculine one at La Oliva. The church was built in 1750, but with various subsequent reconstruction work and enlargements, most recently in 1887.

After finishing the sketch I retired to a bar for a very welcome cup of coffee. According to the barman there were no shops open that afternoon anywhere in Tetir, so I would be unable to get provisions and would have to make do with what I had until I reached Fustes the following afternoon. This was a real blow, although I did still have an orange, a few bits of bread and slices of salami, some chocolate and three small cartons of fruit juice. Fortunately, I was able to buy a litre and a half of water at the bar. With the small amount of water that I still had, I hoped this would be enough to get me to Fustes. My best option was to make the most of the pleasantly cool conditions, and cover as much of the remaining distance as possible in the three hours or so that remained of the day. At least I had eaten well at lunchtime, and I would eat well again once I reached Fustes.

Tetir is expanding rapidly with new houses springing up all around. On the way out of the village there was a scrubby area where a male Spectacled Warbler was singing in the evening sun. His mate rattled her alarm from the deep cover of a cactus thicket, where perhaps she had a nest. Further down the road the unmistakeable yowl of Peacocks came from the settlement of La Asomada, a sound that I subsequently heard in various places. They seem to have become a new 'must have' for an aspiring family. To my right, a pair of Ravens patrolled the sheer cliffs of Castillejo Chico and Castillejo Grande. A view to the south opened out in front of me, revealing a wide and very barren plain with mountains beyond. This was the terrain I would be walking through the following day.

It was time to find somewhere to camp. I crossed the busy main road between Puerto del Rosario and Casillas del Angel, and continued south towards Triquivijate. Just beyond the main road there was a bridge over the Barranco de Rio Cabras, with a trickle of water snaking between tamarisks. Camping in the base of the barranco was a possibility, but there would be flies by the water, and quite a lot of traffic noise from the main road. For some reason I was confident that I would find somewhere further on, even though the light was beginning to fade and my strength was ebbing away. My confidence was in complete contrast to the previous day, when I had seriously thought that my walk was going to end in early defeat.

I was going slowly now as I was very tired, but I managed to drag myself along the road for another three kilometres, constantly looking for somewhere hidden from view where it would be possible to pitch the tent. There were houses scattered all down the road. Finally, just as I was at the end of my strength, I found a suitable flat area hidden from view behind a series of earth banks. The light was fading rapidly by the time I had the tent up and could finally relax and reflect on the day. I had covered twenty-three kilometres, including a challenging mountain section, and my feet certainly knew it: the blisters had both multiplied and grown larger as a result of all the heavy pounding they had taken. But after coming very close to giving up completely, today had been a triumphant one and the fear and uncertainty of yesterday had been completely banished. I congratulated myself on a job well done. It was completely dark by the time I had eaten a bit of food. Somewhat inevitably, a Stone-curlew began calling just outside the tent. I turned off the torch and lay back in my sleeping bag to listen to 'In Rainbows'. All was well with my world.

★

A Lesser Short-toed Lark singing high over the tent just before seven o'clock had the same shattering effect as an alarm going off, and I was up and busy in seconds. I had intended to wake before dawn but had somehow slept too long. Now I wanted to get going as soon as possible as there was a long way to go. I noticed that the weather had changed: there was an easterly breeze and low cloud covered the ridges to the south and east. These conditions sometimes bring migrant birds to the island, drifting east from their normal route up the African coast. I hoped to see a few during the day's walk.

The plan was to head south down the road to Triquivijate, about five kilometres away. It would probably be possible to get breakfast of sorts there, an important consideration given my shortage of supplies. Then I would head down the new road that leads east from there to Fustes, about another twelve kilometres away. I was reluctant to walk on the road as it would certainly be no adventure to do so and traffic always detracts from a walk, but I knew it was the sensible thing to do given my lack of provisions and blistered feet.

It was a steady uphill slog, but at least there were a few flowers beside the road where water evidently runs during wet weather. Amongst them were two vetches I hadn't seen before in the island. One was the familiar Common Vetch *Vicia sativa* with its two-tone red and purple flowers, a plant that grows in my garden back home, but which I was surprised to see here in this barren place. Growing with it was the purple-blue flowered *Vicia benghalensis*. There were also two very flattened Hedgehogs on the road. The species that occurs on the island is the Algerian or Vagrant Hedgehog, which is the African equivalent of the familiar European Hedgehog. It is a little paler in colour, and does not hibernate. In Europe it is found only along the south-east coast of Spain, the Balearic Islands and parts of the south coast of France. Like its European cousin, it is rarely seen because of its nocturnal habits.

After about two kilometres I came to the large stand of Giant Reed *Arundo donax* by La Rosa del Taro, which is several hectares in extent. It is a surprising sight here in such a barren landscape. Any wetland habitats are important for wildlife on this parched island, but sadly this one is very often completely dry, as indeed it was today. There were lots of noisy sparrows but not much else. At least a walk through the towering reeds had novelty value, so I slipped my rucksack off and took the opportunity to lose myself in its curtain of greenery for a few minutes.

Just beyond the reed bed I made an impulsive decision and left the road for the ridge on my left. I would forgo the coffee in Triquivijate and the steady but easy plod of the flat road surface, and walk the hard way to Fustes after all, along the oddly named Cuchillo de Palomares (Dovecot Ridge). It would be a 300m climb from the road to the peak of Rosa del Taro, then a long ridge walk and finally a stretch of flat coastal plain. My remaining supplies were a litre of water, two cartons of fruit juice, an orange, and a slab of chocolate. I hoped this would be enough to see me through.

I put my hat on, smeared sun cream on my hands and face and headed off up the slope. With the heavy pack it was hard work, but it was not dangerous and I made steady progress. A magnificent view of the central plain opened up to the south. I repeated the 'slowly, gently, onwards' mantra as I climbed, and reflected on my decision to leave the road. I could have taken the easy route, but being here was about living my dream, and that required me to take the hard routes and accept the hardships that came with it. I was here for adventure and excitement, not simply to make it from A to B. It had to be the right decision.

About two-thirds of the way up I sat down at a rock outcrop to admire the view. A Hoopoe flew just below me, a flash of black and white barred wings as it passed. It had something in its beak, so must have had a nest somewhere nearby. I picked up my rucksack and carried on up the hill. The vegetation was very sparse, but sticking out of the soil were the exposed tops of the big onion-like bulbs of the squill *Drimia maritima*. These grow in clumps all over the island, and evidently produce showy flower spikes during dry periods, drawing on the nutrients and moisture stored away in their bulbs, but it is difficult to see how they are pollinated at such times as there would be virtually no insects at that time of the year.

As I climbed higher, spectacular views opened up on either side. Then, just below the summit there was a flat area where I was surprised to see a neat, continuous circle of small stones enclosing a space about nine metres across. Near the centre, but not actually in the centre, there were eight larger stones that appear once to have been balanced on each other but were now scattered across the soil. The outer ring of stones was crusted with lichens, indicating that they had not moved for decades, whereas the upward-facing surfaces of several of the central stones were bare, so must recently have been moved. Looking up, I noticed that from here you can see virtually the whole of the island: to the north the plains around Casillas del Angel with Muda and Aceitunal beyond, and a distant line in the haze representing Lanzarote. To the south the whole of the central plain was laid out below, with the mountains along the south coast beyond (Plate 8). It was the most magnificent view of the island imaginable. Suddenly, I found myself wondering whether the location of this circle was significant. But what could it have been for?

Could it be an old wrestling ring? From the descriptions of pre-Conquest islanders' wrestling, it is known that circles enclosed by stones were set aside outside their villages for this purpose. They were used to settle disputes between tribes but also as part of ceremonial and festive occasions. Present day bouts are fought in a ring nine or ten metres across, the same dimensions as this one, and it would not be surprising if the original rings were the same size. And looking carefully at the enclosed area, I noticed that all small surface stones had been removed from inside the 'ring'. It is difficult to think of a good reason why present-day people would climb up here for a wrestling match, but what finer setting could there be for festive bouts than on top of the one mountain with the best view of the whole island? It would certainly have added drama to whatever event was being held. Perhaps the site was established by pre-Conquest islanders, or perhaps by the early Berber settlers. In the archaeological museum at Betancuria there is a picture of the pre-Conquest Jandía wall, which was built by the pre-Conquest islanders to separate the tribes living on Jandía from the main part of the island. In the photograph there is a similar circular structure set against the wall, so perhaps a pre-Conquest origin is not too far-fetched. Whatever the origin of the circle, it is certainly intriguing, and well worth the climb for the view.

★

After taking all this in for a while, I climbed the last bit of Rosa del Taro. At the top there was a surprising amount of clutter: a trig point, a flag pole, the remains of some kind of mast, and a padlock lying on the ground complete with key! I could now see along the ridge, and Fustes was just visible on the coast in the distance. The ridge was rather spikier than I had hoped, and it looked as though the walk would be distinctly energy-sapping. To my left there were views into the deep valley of Barranco de Jenejey, which has a huge cement works at its eastern end. I was glad not to have walked to Fustes through that valley, which was one of the options I had earlier considered.

As I started out along the ridge, a Buzzard mewed and circled past me. Then I came upon a sheep with two very young lambs. The sheep was the dusty reddish colour of its surroundings, but the lambs were the purest white, too young yet to have taken on the colour of the island. A little farther on two small birds flushed in front of me, and from the calls I knew they were Tawny Pipits, scarce migrants to the Canary Islands, and the first migrants of interest during the walk.

Clearly the weather conditions were indeed bringing birds over from their intended route up the west coast of Africa.

It was an exhilarating place to walk, with wonderful expansive views on either side to other sheer-sided ridges. In front, the airport drew slowly nearer, with plane loads of tourists arriving and departing as the day wore on. An Armas ferry left the harbour in Puerto and headed south, bound for Gran Canaria. Much closer to hand, an Egyptian Vulture flew along the ridge, giving me good views. For the vulture it was an effortless glide of a few minutes between one end of the ridge and the other, but for me it was a tough, two hour slog. There was no hint of a path and the ground was incredibly rocky, so I had to pick my way carefully along it. It seemed to go on for ever.

At last I stood on the easternmost rock outcrop that marked the end of the ridge. I took out my map and surveyed the plain below, trying to work out the best route, but there was no obvious set of tracks going the right way. In the end I simply took a compass bearing and headed straight towards Fustes. I dropped down from the ridge and was on easy ground at last: now it would be level all the way to the café, or so it seemed. But there were still five kilometres of hot work ahead, and my water supply was running low. When I came to the Barranco de Muley, I had to find a way down what was effectively a cliff. Having done so, I sat on a rock to drink the very last of my water. Another Hoopoe gave lovely close views and two Willow Warblers flitted about in the scrub. I rested here for a few minutes, then set off on the last leg of the journey to Fustes.

There were now just a few kilometres to go, with no major obstacles to contend with, but I was really tired, and I was limping because of my sore feet. It was hot, and I knew I was beginning to get dehydrated. However, the lure of a comfortable apartment and unlimited supplies of water (and food) was very strong; it was simply a matter of carrying on as best I could for another hour or so. The terrain was uninteresting: the eastern coastal plain is the least scenic part of the whole island. Birds were almost non-existent and progress was largely head down, although a familiar chortle made me look up at one point to watch two lovely Black-bellied Sandgrouse that flew up in front of me, giving my flagging spirits a momentary lift.

I finally came to the coast road, and after following this south for a while I could at last see Fustes laid out before. The last kilometre was really hard. Finally, there were just eleven more telegraph poles between me and the edge of town, and I counted them down as I struggled to complete the walk. They seemed to go on for ever, and I had to stop and recount them twice when the number in front of me appeared to be going up!

Then, suddenly, I was sitting in a comfy chair in a cool bar watching the racing from Cheltenham with a bottle of cold fizzy water and a bag of salted peanuts. I had made it but now I was utterly exhausted. Over the following hour I drank several more bottles of water, trying not to hobble too pitifully to the bar each time I needed another. I was in the completely alien environment of an Irish bar, waiting for my body to recover enough to head back out in search of my hotel. The contrast between the world I had lived in for the last three days and the one I now found myself in could not have been greater.

TWELVE

Caleta de Fustes

In the week or so before leaving home I had spent quite a lot of time deciding on accommodation. I had wanted to stay in the centre of the island, but try as I might I simply couldn't find any accommodation away from the coast. Nevertheless, I was determined to rest somewhere halfway down the main part of the island, so I eventually had no option but to divert to the coast at Caleta de Fustes. Like Corralejo, this resort is also heavily dependent on British and Irish tourists.

Initially I was worried that there would be no accommodation available, because March is high season on the island. Northern Europe is still cold at that time of year, and after six months of dark, wet days, the lure of warm sunshine is almost irresistible. Worse still, my planned stay in Fustes was over the weekend before Easter, which is just about the busiest time of all. Nevertheless, there was still a huge variety of places in Fustes with rooms available in the week before my departure, which seemed to be a bad sign for the tourist industry on the island. That being the case, it was really difficult to decide where to stay. I didn't want anything too expensive, just somewhere I could laze around for a day or so to recover. After spending many hours comparing the different options I took a virtual pin and booked two nights in the La Tahona Apartments. Having done so, I checked out a couple of review sites and it appeared that I had just booked myself into the place with some of the worst reviews in Fustes. I made a mental note to check reviews in future before booking accommodation on the internet next time. That said, I found the accommodation more than adequate, and I now realise that it is not always wise to take internet reviews too seriously.

It was with some misgiving, therefore, that I dragged myself out of the bar where I had been recovering from my walk and began looking for the apartment. Thankfully it was only a few hundred metres away, and I found it without any difficulty. Even so, it was hard-going because my feet were terribly sore now and having made it to Fustes, my body just wanted to stop. Thankfully, check-in was straightforward and within moments of arriving I was shown to my room.

After throwing the pack onto my bed, the first thing I did was take my socks and shoes off. Oh the bliss of it! Now that sanctuary had been reached, I realised how utterly focussed I had been on the logistics of the journey, the effort of the walk and my surroundings. After just three nights of rough camping the facilities now available to me seemed like untold luxury. For example, there was a mirror, which I looked into as though it was an entirely new invention. Thankfully, the sight that greeted me was not too terrifying: I needed a shave, but the combination of hat and sun cream had prevented sunburn. But my poor feet were in a bad way with huge bulging blisters over a large part of the soles and toes. Having carefully washed them, all I wanted to do was to sit in a cool spot and do nothing, but before I could relax I had to get some provisions. I put on my sandals, which to my relief were a little more comfortable than the boots, and hobbled down to the nearest supermarket.

Having made it back to the apartment, I could finally relax in the knowledge that I didn't have to do anything much for a day or so. There was no objective now other than to rest and allow my body to recover. I had a belated lunch of all kinds of goodies, then the unimaginable luxury of a hot bath. After that I tried to have a shave, but for some strange reason I could not get my shaving cream to lather at all. No wonder, it turned out that I had been using fungicide cream rather than shaving foam!

A couple of hours later my wife phoned. It was wonderful to hear from her. I have no idea what we talked about, but afterwards I felt as though I was back in the real world. After a gloriously comfortable night's sleep, having eaten a huge quantity of breakfast in the canteen, I bought a red Lanzarote wine and yet more goodies, and spent much of the day sitting in the warm sun on the patio slowly drinking the wine, nibbling treats and enjoying the rest. The contrast with the previous day, when I had had a rudimentary breakfast and walked for six and a half hours on blistered feet, could not have been greater.

Just in front of the patio there was a large, spreading acacia tree that looked as though it might attract a few birds. Every now and then I glanced up to see whether anything had appeared. Apart from an occasional Spanish Sparrow or Collared Dove, the only visitors were two Chiffchaffs (Box 14), one of which spent a fair proportion of the day flitting through the leaves in search of insects.

★

It was late morning when I headed into Fustes, hiding my limp as much as possible. I hobbled nonchalantly past the Goose and Firkin pub and innumerable trinket shops in search of a new notebook as my first one was rapidly filling with scribbled notes. The only shop in town that sold notebooks of any kind charged an exorbitant tourist price, but I had no choice but to pay up. Then I headed out along the main street towards a rather good-looking café I had noticed on the way into Fustes. The Pastelería Pantarajal is definitely to be recommended. It is beyond the tourist centre and is much frequented by local business men and women. It serves real coffee and great cream cakes.

Almost everything else in Fustes is oriented towards tourists. Nearly all the bars have English names, show live English football or horse racing, are staffed primarily by British or Irish people, and offer cooked breakfast and Guinness. There is a large sign advertising the British Surgery for anyone who gets too much sun, reacts badly to the local food, or breaks a limb whilst drunk. Clearly there are a lot of people who want guaranteed sunshine but otherwise want everything to be the same as it is at home. If so, then this is the place for them. I must admit I struggle with it all. When I go on holiday I want to feel that I am somewhere different, to sample local foods and generally experience new things. But there is one big advantage of somewhere like Fustes – there is no language problem. Almost everybody speaks near perfect English. Life is incredibly straightforward. For me, it was like popping back home for a couple of days – except of course that I didn't know anybody.

★

Back at the hotel, I noticed that the desk calendar at reception had a picture of a displaying male Houbara Bustard on it. Fittingly enough, this was the picture for March. I explained that I had studied the bird in Fuerteventura many years before, and asked if I could have a look at it. Each month had a photograph of the island or one of its characteristic species. I asked how I might get hold of a copy to take home with me. The charming staff explained that it was not for sale but had been provided by the island council, but seeing how much I wanted one, they simply gave it to me. It now has pride of place in my office.

★

On my second morning in Fustes I woke up wondering what to do with the day. One day of lazing around was great but, despite my feet, the thought of a second day doing not very much was not appealing. So I hired an inexpensive but rather battered Toyota Yaris so that I could visit some of my favourite places that I would not be walking through. Then I rang Derek Bradbury, who has been providing a spring bird guiding service based up at Cotillo on the north-west coast for some years. I arranged to meet him that evening.

★

Back in the early 1980s, Fustes was the smelly place you drove past on your way south from the airport, definitely with the windows up, and preferably at high speed. Quite why the place smelt so I don't know, but smell it certainly did. Needless to say, in those days there was no tourist development, just a few low buildings clustered around the cove of black volcanic rock, with its eighteenth-century defence tower. The cove is unique on the east coast of the island in that it faces south, and is therefore sheltered from the predominant northerly winds. For this reason it was an important anchorage in the days of sailing ships. The circular tower, which is very similar to the one at Cotillo, was built around 1740 to provide a defendable retreat when pirates visited the island. These pirates were, as often as not, British. How ironic then that the settlement is now a favourite resort for British holiday makers.

A tourist guide to the island published in 1980 wisely makes no mention at all of Fustes, despite the interesting defence tower. By the end of that decade, though, the smell had been vanquished, and a small resort had grown out as far inland as the coast road. At that time there was a thick hedge separating the road from the villas beyond, and this was a major attraction for migrant birds. I once found a Nightingale lurking in the dense undergrowth for example. It was one of the few patches of cover on the whole of the east coast of the island in those days. Now there are landscaped gardens throughout the resort and the migrants are more scattered and harder to find.

Until fairly recently, the road south from Fustes crossed a barren plain. However, in the last decade the pace of development has been staggering, and there are golf courses and hotels for a fair part of the three kilometres to Las Salinas, where there is a salt-works museum. The saltpans themselves, the Salinas del Carmen, are quite small and not particularly impressive, except for the whale whose skeleton stands guard over the site, held aloft on metal supports.

Just south of the saltpans is a cluster of houses known as Casas de las Salinas, where there is a good restaurant. The main road swings inland, but there is a driveable track south through Las Salinas which continues for another kilometre or so to the mouth of the Barranco de la Torre. In fact, until about 1980, this was the end of the coast road.

Barranco de la Torre is one of the main barrancos on the east coast of the island, and takes its name from the 'tower' where the barranco meets the sea. It is probably the longest barranco on the island, rising in the Betancuria mountains above Antigua, and is known as Barranco de Antigua for most of its length. The tower, after which the lower end of this great barranco is named, is a major limekiln. In the past lime was produced here as well as in many other places by burning the calcareous crust that covers so much of the island.

Box 14 *Chiffchaffs in Fuerteventura*

The Chiffchaff is a small greenish warbler, very similar to the Willow Warbler except for its completely different song and, to the practised ear, a slightly different call. In all the Canary Islands to the west of Fuerteventura there is a resident form of the Chiffchaff that is now considered to be a separate species – the Canary Island Chiffchaff. A race of this bird used to occur in Lanzarote, where it is now extinct, but there appears to be no evidence that it ever occurred in Fuerteventura. The Chiffchaffs that are seen in Fuerteventura are winter visitors from the colder parts of Europe. They are inconspicuous little birds, but if you look hard enough you will find them anywhere where there are bushes and trees. As they are on their winter quarters here you don't often hear them singing, but from time to time they sing a few bars of the typical chiff-chaff song that gives the bird its name, and just occasionally a snatch of a rather more complex, very different song that identifies the bird as an Iberian Chiffchaff, which is now also considered to be a species in its own right.

In mid-March, many British Chiffchaffs have already returned to the woods and hedgerows where the blossom of wild cherry is already bursting. Not so thirty years ago, when there was no point looking for them until the last few days of March, but as the world slowly warms, they fly north earlier. It seemed likely that the birds in the tree outside my room were from much further north or east where the land was still in the grip of winter. In Latvia, perhaps, where the Chiffchaff's song is not heard until the end of April, or the hillsides of northern Norway, where the ground would still be covered with a metre of snow. No wonder it had chosen to stay here for a while longer, eating aphids in a sunny acacia tree.

In contrast to the Chiffchaff, the Willow Warbler is a passage migrant on the island. It spends the winter in the forests and savannas of sub-Saharan Africa, as far south as Cape Town.

Thirteen

Pájara and Ajuy

The main road south from Fustes turns inland at the saltpans and after a few kilometres crosses the impressive rocky chasm that is the Barranco de la Torre. From there, most traffic heads south-west towards the tourist resorts of Jandía. I drove due west across the flatlands that dominate the centre of the island. This part of the island was once a wide expanse of nothing: just a monotonous empty plain frequented only by goatherds and bustards. Now it has a rash of houses: sparsely scattered, but sufficient to alter its character.

On the western margin of this central plain lies Tuineje, the administrative centre of this part of the island. Despite its relative importance, Tuineje is unremarkable and has changed little over the years. Indeed its lack of development now appears to be its main selling point. From Tuineje I took the road to the north-west that climbs steeply into the southern end of the Betancuria hills before dropping towards the villages of Toto and Pájara.

★

The tidy village of Pájara has always been a green oasis with cool and pleasant public spaces. This is in stark contrast to Tuineje and the dusty villages of the central plains. At the eastern end of the village is a leafy public park and the centre has stately conifers, eucalyptus and other evergreen trees. The eastern approach is lined with more stately palms than most of the towns on the island, and at the roundabout that marks the entrance to the village, there is a rather good sculpture of a goatherd milking goats.

Pájara is the administrative centre not only for the south-western end of the island, but for the whole of Jandía as well. In the days when the Jandía peninsula was accessible only along an interminable dirt track, this must have been something of a handicap, but today it brings the benefit of substantial income from the seemingly never-ending development. The town has recently benefitted from considerable investment in its village space. There is now a nicely landscaped path beside the town's barranco, a pleasant place for a leisurely stroll. Its exotic shrubs, and the massed red and orange flowers of Bougainvillea provide a wonderful antidote to the barrenness of the surrounding countryside.

This lush spot is a good place for migrant birds, and in winter there are small numbers of birds such as Blackcap and Chiffchaff, and perhaps the odd Robin and Song Thrush. The bird I most associate with Pájara is the Sardinian Warbler, which is particularly common here at any time of the year. The male is one of the more striking warblers. It has an overall grey colour, but with a black head and vividly contrasting white throat, red eye and red legs. It is quite skulking and would be hard to spot if it wasn't such a noisy bird. It gives itself away with its strident calls and scratchy but not un-tuneful song. Within Fuerteventura, it is restricted to areas that have historically been well-vegetated. These include both the mountain valleys and wooded areas between Pájara and Betancuria, and the larger barrancos with significant Tamarisk thickets in the southern half of the island. It seems to be remarkably sedentary, and is hardly ever seen in places such as Antigua, La Oliva or Corralejo. I did once see one in an isolated patch of cover west of La Oliva, but this may have been a wintering bird from Europe, as some European birds are migratory.

During the Honeyguide wildlife holiday that I led to the island in 1996, we watched a male Sardinian Warbler attacking its image in a car wing mirror in the main street in Pájara. It was so intent on this that it took no notice of us and we were able to get very close to it.

Today, as I sat on a bench to enjoy the cool shade of the shrubs along the barranco walk, a pair was busy feeding its recently fledged young. One young bird with a very stumpy tail was hopping about on the ground only a few metres from where I was sitting. At one point the male brought what appeared to be a Greenish Blacktip butterfly. In fact, the area around the barranco is a good spot for butterflies, and on a good day quite a few species can be seen here.

One interesting butterfly that can sometimes be seen here is the Geranium Bronze, an undistinguished looking small brown butterfly with short 'tails' on its hind wings. It comes from South Africa where, as its name suggests, its caterpillars eat Pelargoniums and Geraniums. The species was first noticed around the Mediterranean in the early 1990s, and it is now found in many areas. It first made an appearance in Fuerteventura in the early years of the twenty-first century, and seems now to be established in tourist resorts and some of the larger villages.

★

Whilst the trees and public spaces certainly make Pájara an attractive place to visit, the star attraction of the village is undoubtedly the church of Nuestra Señora de la Regla, which takes pride of place in the centre of the village. By Fuerteventuran standards it is quite large, and there are two aisles. But the feature that makes it remarkable, and indeed unique, is the fabulous decoration on the west front of the north aisle. The guidebooks describe the designs as 'Aztec-like', but I think this is mistaken. Admittedly, the geometrical patterns and curious lobed motifs are very strange, but they bear no resemblance to the angular figures and evil, blood-thirsty monsters that are the central themes of Aztec carving (Box 15).

There are a few bars in Pájara that provide food, but for a really good meal it is worth driving on to the west coast at Ajuy, where there are several good fish restaurants. After a pleasant morning in Pájara, I drove out of the village with its cool shady streets and incessant chirping sparrows. The desert landscape beyond the village seemed more barren than ever.

★

A road off to the right about halfway between Pájara and Ajuy takes you to the little settlement of Buen Paso. Beyond it there is a wonderful view of the palm-filled Barranco de las Peñitas. The barranco continues through a deep gorge from here towards Vega de Rio Palmas. A rough track through the gorge passes a humble little hermitage overlooking permanent rock pools. You could walk to it from here, as there is a rocky path up the barranco, but a more popular option is to walk from the other end.

Beyond the turning to Buen Paso, the road follows what is perhaps the most impressive barranco on the island, with stately palm trees and a few abandoned houses. It was here, amongst the tall palms, that the naturalist David Bannerman enjoyed camping during his bird collecting expedition in 1913. Ajuy itself is a little fishing village clustered around a cove at the end of the barranco. The small gritty beach is open to the full force of the Atlantic. Sun-worshipers paddle daringly at the edge of the waves and a few surfers try their luck in the swell.

Apart from the fish restaurants, the other main attraction at Ajuy, and worth a journey in its own right, is the short walk along the cliffs to the sea caves. Not only is this one of the most impressive walks on the island, it is also well laid out and therefore accessible to tourists. Although it is now simply a walk to a sea cave, the reason for the path being here in the first place is that it was once the landing point for boats on this part of the west coast. David Bannerman gives a vivid account of a precarious landing here in a small boat in 1913, having made the journey south from Cotillo.

"We caught the breeze squarely as we rounded a rocky headland and the boats fairly plunged through the waves into a good-sized bay surrounded by cliffs, which were literally honeycombed with caves...The heavy swell rolling into the bay tossed our frail craft about like shells, and as the waves roared into the caves or came thundering against the cliffs, clouds of spray were sent fifty feet into the air. There appeared to be no place on the face of the cliff better than another at which a landing could be accomplished, but our sturdy Canarian sailors managed their craft with extraordinary skill, and deftly ran the boat alongside a shelving ledge. The rise and fall of the swell at this point seemed to be accentuated with our proximity to the cliff; now the boat was poised on the crest of a wave and we rose far above our former level, only to fall again as the swell rolled onward into the bay. As we rose again, one of the boatmen leapt at the ledge, choosing the exact moment when the boat seemed to be neither rising nor falling, and a moment afterwards he was left clinging to the precipitous rocks far above us. As the next wave heaved us up, a second sailor sprang for the cliff, and, one by one, my wife, her sister, and the rest of us, made our leap for the land, being caught be the two sailors on shore and hauled by our wrists up the side of the cliff to a place of comparative safety. It was certainly the most exciting landing I had ever made, and even when left to our own devices, we had to crawl like cats on all fours from ledge to ledge, bumping our heads on the oft-overhanging rocks."

The path to the old landing place is indeed quite steep, but now has proper steps and wooden railings for a secure descent. The metal rings that would have been used to secure the boats' ropes are still in place on the slight promontory at the mouth of the cave, and there are remains of a pulley system that must have been used to unload cargo.

The views across the bay are splendid enough, but the sea cave is magnificent. There are steps down to the floor of the cave itself. With care (a torch is ideal), you can then clamber into the cave over the rubble and rocks that have fallen from the roof, until you come to a small hole where a shaft of light enters the side of the cave. From here you can look out to the surf crashing towards you just below, a thrilling experience. It is also possible to clamber over rocks into the adjacent flat-bottomed cave. This cave narrows gently and you can walk as far as you dare into the darkness. Then, turning round, there is the seething blue sea and white spume filling the mouth of the cave behind, and the sound of the roaring sea pervades everything.

There is a strange difference between the two caves: the first has a floor of sea-worn pebbles right to the upper end, which is well above the current reach of the sea, whereas the second has a floor of bare, jagged rock. Could it be that a tsunami deposited the pebbles in the first cave at some time in the past when the second cave was not yet opened to the sea?

★

About a kilometre or so further north is the Peña Horadada (literally the perforated rock), which stands at the sea's edge at the mouth of the Barranco de la Madre del Agua. This natural arch was presumably once the mouth of another sea cave. It is possible to stand under the arch, although I found this a slightly unnerving experience. Once under the arch you can see how thin and precarious it is, and the force of the breakers is sometimes sufficient to spill quantities of water into the rock pool that lies under it.

There is no marked path to this natural feature, but if you want a rugged walk, this is as good a landmark to head for as any. The barranco itself is full of tamarisk *Tamarix canariensis*, and therefore Sardinian Warblers, and the rock is certainly impressive.

Box 15   *The façade of Pájara church (Plate 7)*

The carvings on the façade of the church of Nuestra Señora de la Regla at Pájara are so unusual that it is perhaps no surprise that people have sought a pre-Columbian American origin for them. No doubt the feathered head-dresses also suggested such an origin. However, head-dresses in Aztec carving are fantastical, many layered structures, as they were in earlier pre-Columbian civilisations. Having visited several exhibitions of Aztec and Mayan sculpture and art, and trawled through many of the standard texts on the subject,

I can find nothing that remotely resembles the decorations at Pájara. Also, there are other churches in Western Europe with diamond and lozenge patterns. One thinks, for example, of the great Norman cathedral in Durham.

The general layout of the façade is classical in origin, with two pilasters on either side of a round arched doorway, surmounted by a triangle (mimicking the pediment of classical architecture) within which is a single round window. It is fairly grand by Canary Island standards, but not exceptionally so. The front of the church of El Salvador in Las Palmas, for example, is not dissimilar in layout. The decoration, however, is an extraordinary mixture of figures and abstract designs. With the exception of the much less extensive decorations on the private chapel of the military governor at La Oliva, it is quite unlike anything to be seen anywhere else. No part of the façade is without decoration: where there is no other ornamentation, the space is filled with diamonds and lozenge patterns.

The most unusual decoration is in two bands. The course of stones above the top of the door (the architrave) has two crudely carved snakes in the centre that are swallowing their tails, a common enough motive in mediaeval European churches. But the remainder of the course is composed of highly unusual geometrical designs that have been interpreted as sun or moon symbols, though they are perhaps just abstract patterns. On either side of the snakes there are two objects with circular centres that sprout lobed projections. Outside these, there are two more blocks on either side with sunflower-like designs.

The course of stones above the triangle and immediately below the roof, are even more fantastical. One of the figures is clearly a dove, and there are two beasts that might be lions either side of a central winged figure, presumably Christ, which is topped by a small cross. Either side of Christ there are human heads. The strangest thing is that both Christ and the two heads either side have what appear to be feathered head-dresses. On either side of this group of figures and heads there are very strange objects that appear to be six-headed beasts, each of which has a bulbous tongue sticking out from its open mouth. Another very unusual feature is that, at some stage, the whole of this façade was washed with a pale blue paint, traces of which can still be seen.

We will probably never know what really influenced the carvings, but what is certain is that an imaginative mind was at work here in Pájara. We should not forget that Fuerteventura was quite an isolated place in the seventeenth century, especially up here in the hills. A local sculptor instructed to cover the façade with carvings would probably have had very little to work from, and was perhaps reliant on images in a few books, what little ornament he could see on other churches nearby, and his own vivid imagination. With this in mind, it is worth noting that the doorway of the old cathedral a few miles away in Betancuria might just have provided at least some of the inspiration for these carvings (more on this later). Perhaps the art represented here is best celebrated as a unique Fuerteventuran cultural flowering.

Fourteen

Cotillo

On my last evening in Fustes I drove north to meet Derek Bradbury in Cotillo, or more correctly El Cotillo, though nobody ever seems to call it that. Firstly, though, I wanted to spend the last hours of the day looking for Houbaras. The coastal plain south of Cotillo is one of the most reliable places on the island, and indeed in the world, to see these magnificent birds.

Although most birdwatchers do manage to see at least one or two Houbaras during a week on the island, some fail to do so despite hours of searching. Given that I only managed to see one in the first two weeks of my studies, this is not entirely surprising: their numbers are quite low and they are expert at avoiding detection. However, by following a few rules they are much easier to find.

The first, rather obvious rule is to find good habitat: flattish ground with scattered shrubs, no buildings, and preferably with some cultivated land nearby. The second rule is to look for them either just after dawn or in the cool evening hours. Houbaras are largely inactive during the middle part of the day, and it is no easy matter to spot one that is sitting tight. If you are on foot, the third rule is to find a suitable vantage point and stay there for at least fifteen to twenty minutes, scanning carefully with binoculars. When you first appear, the bustards will often either squat or hide. After watching you for some minutes, they usually become active again and hence much more visible. However, the best thing to do is undoubtedly to hire a vehicle and find somewhere where you can scan from inside it. Houbaras virtually ignore vehicles and there is a much better chance of getting close views of birds behaving naturally from a vehicle. But please, do stick to roads and main tracks and don't be tempted to drive your vehicle over the plains because this disturbs the birds and could well destroy nests of Houbaras and other birds.

Although the plain south of Cotillo is as good as any to look for Houbaras from the car, even here they cannot be guaranteed. On this occasion, I initially took the wrong track that went up to some houses, but having rejoined the main coastal track I stopped in a suitable open spot and almost immediately had very close views of three splendid adult males unhurriedly crossing in front of the car and feeding nearby. After watching them for some time I drove on over a slight rise and stopped again. My luck was definitely in. This time I immediately saw a female with three half-grown young. Her behaviour was so utterly different to the males, dashing here and there in search of food with her offspring running along behind and trying to keep up with her. I watched them for about a quarter of an hour until they disappeared over a ridge.

I met Derek and his wife at the restaurant El Rustico in Cotillo. It was good food, and a real treat to eat in company for a change, and to hear Derek's story. After triple heart bypass surgery he retired from the hotel business and came out to Fuerteventura for three months to recuperate back in the late 1970s. They fell in love with the place and have been back for three months every year since. The dry climate is beneficial to his health, and he has become a dedicated Houbara watcher. He knows all the spots around Cotillo where males display, and keeps track of them through the display season (December-March) and shows them to visiting birdwatchers.

★

Until very recently Cotillo was unique in two ways: it was the only settlement of any size on the west coast of the island, and it was also the only village with a superb beach that had not been developed for mass tourism. During the 1970s, several highly eccentric modernistic villas (follies might be a better word) were built in the dunes north of the town. For some years, these whitewashed creations with their stair turrets and arches were virtually the only sign that Cotillo had been noticed by foreigners. Back then, they were in such contrast to the huddled workaday houses of the nearby fishing village that they seemed almost magical. There is also a long established hippy community, drawn to the area by the world-class windsurfing beaches immediately to the south and along the north coast towards Corralejo.

As recently as 2004, my daughters and I had a wonderfully relaxing holiday in the Maravilla apartments, a small development on the shore just north of Cotillo. It was the perfect spot for the girls. There was a small swimming pool in the courtyard just outside the back door, and a garden terrace where we could sit in the hot sun and read, looking out over the beautiful white sand beach. Sadly, the first hotel was then being built on the northern edge of the village.

The only surprise is that it has taken so long for tourism to take off here. Fortunately, the scale of development has so far been modest, and the village still retains much of its character. And of course development is not all bad. The recent arrival of a good supermarket and a French patisserie are very welcome. Nevertheless, a crash in the tourist industry is probably the only thing that can save Cotillo from losing its character, as has happened in so many places around the world in recent decades.

★

To the north of Cotillo the coast is formed of lava that flowed from the nearby volcanoes. The rock is black and as jagged as the day it solidified. Between fingers of lava there are shallow bays filled with sand of the purest, dazzling white that colours the water deep blue or turquoise, depending on the state of the tide and the light. This being the west coast, towering waves throw white spume up as they crash against the reef edge, creating a constant, exciting roar. And best of all there are sheltered bays where the reef all but completely subdues the waves, which remain calm and perfect for swimming. I am not a keen swimmer, but even I can't resist a dip here. This shore is also as good as anywhere on the island for wading birds such as Turnstone, Sanderling, Kentish and Ringed Plovers and Whimbrel. On one memorable occasion, I swam to within a few feet of a group of Sanderlings, getting a crab's eye view of them picking for invertebrates at the water's edge. Presumably a swimming human is not recognisable as such.

The plants growing in the wind-blown sand just to landward of the beach itself include a number of interesting species, the rarest and most beautiful of which is the endemic sea-lavender *Limonium papillatum*, with its sprawling greyish branches covered in frothy pink flowers after winter rains (Plate 9). Another interesting plant is *Astydamia latifolia*, a yellow-flowered member of the carrot family with large succulent leaves that grows on the coastal rocks.

These sheltered beaches used to be a favourite holiday resort for local families, who traditionally camped here. This seems to have stopped now, presumably because the local authorities have clamped down on the practice. Not so along the south coast, where the locals still camp (more on this later).

★

Four kilometres north of Cotillo, on a platform of lava just above the heavy surf, stands Tostón lighthouse. Cotillo itself was known as Tostón until some time in the twentieth century. The lighthouse is set amongst some of the wildest coastal scenery on the island. It has no less than three towers, all of different ages. The original structure was built in 1897, with a small domed turret to house the light. A somewhat larger tower was added in 1950 and finally the tall, red and white hooped tower that dominates the coast was constructed in 1986. The lighthouse was renovated in 2008/9 and now houses a museum of traditional fishing, which tells how the original inhabitants of the island caught fish hereabouts by poisoning the shallow waters with spurge milk.

Until recently there was just a rough track to the lighthouse, but, as part of the ill-judged drive for development, a metalled road has been constructed, complete with visually intrusive lamp posts. Although it was approved by the local council to enable development, the decision was hotly contested, and work has now stopped. There is certainly bad feeling about the matter amongst a section of the local community. During my visit in 2008, the following words were painted across the road: 'Isla sin ley – solo CORRUPCION – CALLE DE LA VERGUENZA.' (Island without law – only corruption – street of disgrace.) It remains to be seen whether the road will be removed or whether the developers will have their way. If development comes, a place of beauty will have been destroyed, and this will surely be to the detriment of Cotillo in the long run.

★

At the southern end of Cotillo is the 'new' harbour, created a few years back by constructing a massive concrete bulwark between the mainland and the steep islet of Roca de la Mar. This grotesque bastion stands 10m above the quay. It needs to be high to keep out the huge waves that frequently batter this coast, but sadly it ruins the coastal landscape.

Until the bulwark was constructed, the local fishing fleet hauled out in the sheltered bay a few hundred metres further north, in the centre of village. Above this now empty harbour there is a fine statue of a woman gazing out to sea, her right hand sheltering her gaze from the setting sun. The statue is dedicated to the island's fishermen. She is waiting for her menfolk to come home: you sense from her concentration that they are overdue and she is worried that they may have been drowned. Now she waits by an empty harbour to which the fishermen will never return.

In fact, the 'new' harbour must have been the main harbour in past centuries, because it is overlooked by the eighteenth-century defensive tower (Plate 10). This tower is the sister of the one at Fustes and was built at around the same time. Most authors agree that it was built in 1740, although the information provided at the tower today claims it was built in 1700. More information about this fort and other defensive structures on the island is provided in Box 16.

The tower is open to the public, and there is a splendid view from the top. To the south there is a long beach and a bay much frequented these days by kite surfers. Behind the beach, a line of black cliffs points south past the Betancuria mountains towards the more distant mountains of the Jandía peninsular, which are only visible on a clear day. In the other direction is the harbour and Cotillo itself, and close by, just above the harbour, are the towers of several restored lime kilns.

During his visit to the island in 1913, David Bannerman spent several days here. In his book about his travels in the Canary Islands there is a photograph of his camp in the lea of the defensive tower, and another taken from the top of it. And the strange thing is that in this photograph the lime kilns aren't there, so they must have been built after that date. The lime kilns here and further down the coast at Ajuy testify to the fact that lime was being exported to Gran Canaria, presumably to fuel the rapid expansion of the port at Las Palmas, which rather suddenly became a major port in the first half of the twentieth century.

★

Prior to the construction of the new harbour, the Roca de la Mar was an all but inaccessible islet only 200 metres offshore but with virtually sheer cliffs on all sides. It must have been rarely visited by people, and clearly never by goats, because on the flat top a rash of big yellow flowers grew every spring. These were clearly visible from the mainland through binoculars, and it looked as though they might be the flowers of *Pulicaria canariensis*, a rare fleabane that is endemic to Lanzarote and Fuerteventura. To confirm this, it would have been necessary firstly to find a fisherman who was willing to take me over to the islet, assuming the sea conditions were suitable, and then to find a way up the rock to reach the flowers on the top. This was a daunting prospect and one that I never pursued. However, the new harbour structure now butts rudely against the northern end of the rock, and access to confirm the identity of the fleebane is now straightforward. The plants form clumps with bright yellow flowers 3cm across and it is well worth scrambling up the rock to admire them (Plate 11).

Box 16 Defensive buildings

Defensive structures were built in Lanzarote, Gran Canaria and Fuerteventura in three distinct periods. Several forts were built in the days of initial settlement and conquest in the fifteenth century. However, no trace of them has been found so their form is unknown. As well as providing basic shelter from sun, wind and rain, they would also have been constructed as defences against whatever threat the original inhabitants posed. It is just possible that more substantial forts were constructed to defend the early settlers against other potential European invaders, but any such forts would surely have left visible remains. In any case, Fuerteventura was a remote place in those days, and it seems unlikely that this was a serious consideration until later on.

In Fuerteventura there is mention of two early forts. One called Valtarahal was built somewhere on the east side of the island, and the other, Ricorroque, was built in 1404 on the western shore, although again the location was not recorded. Whether the present tower at Cotillo is on the site of this early fort is disputed, but as far as I am aware there is no evidence that it was. In all probability the early forts were simple stone enclosures, which would surely have been sufficient given the vastly superior weapons of the European settlers.

Ricorroque was named after the Norman ship Riche-Roche: it was a group of Norman noblemen who led the conquest of the island. Is it significant, then, that the original name for Cotillo was Puerto del Roque, and that the little settlement one and a half kilometres inland is to this day called El

Roque? And if Cotillo was initially known as Puerto del Roque, it seems likely to me that the original settlement was at El Roque itself. It is on the edge of the *malpais*, in a place where water collects naturally after rain. Hence it is cultivable. To this day it produces figs and other crops, and would have been a site where a permanent settlement could be made whereas Cotillo is on a dry plain and is of use only as a port. If I was going to look for remains of the original fort I would look at El Roque, not at the site of the existing defence tower.

The second period in which defensive buildings were constructed was in the late sixteenth century. Unlike the first forts, these buildings were designed primarily to ward off attack from the sea. The Berbers were on the rampage and were capable of causing serious damage. A good surviving example of a defensive structure built at this time is at Arrecife in Lanzarote, where the Castillo de San Gabriel, constructed in 1590, dominates the offshore reef. On Gran Canaria two defensive towers were built at the anchorages of Las Palmas (completed in 1577) and the Bay of Gando.

There appear to be no surviving buildings from this second period of fortification in Fuerteventura, although it is of course possible that one or more of the forgotten ruins that dot the island (the one on the northern shore at Corralejo, for example?) date from this period. However, it is claimed that the present tower at Cotillo, which was part of the third period of defensive construction, was built on the site of a sixteenth-century tower. This seems likely, since their function was the same (to defend the island against attack from the sea) and it is known that the similar tower at

Gando was built on the ruins of the old one, just a year later than the Cotillo tower was built (1741). By the eighteenth century, the islands' inhabitants were menaced by pirates of various nationalities, including British, French and Berbers.

Surprisingly, although the towers at Gando and Cotillo were constructed at about the same time, their designs are radically different. The Gando tower, which like the one at Cotillo has been restored, has a central pillar to support the heavy stone gun platform, and the staircase is in the main space inside the wall. At Cotillo there is no central pillar, just a simple barrel vaulted roof, and the staircase is within the wall itself. Perhaps their designs were partly influenced by the structure that preceded them. These two basic designs, with modifications, were subsequently used by the Spanish around the Mediterranean, and then copied by the English on the south and east coasts during the early part of the 19th century, when the 'Martello towers' were constructed.

FIFTEEN

On the road again

It was now time to start walking again, but what route should I take from Fustes? My original plan had been to walk virtually due west from here, starting with the ridge of Cuchillete de Buenavista to Antigua, then on through the old capital at Betancuria to the coast at Ajuy. Then I would walk up the Río Palmas Barranco and south through the mountains to Pájara, from where I would head south-east through Tuineje to my next hotel at Las Playitas on the south coast. I had spent years considering the best route, and had finally settled on this as the best way to incorporate all my favourite bits of the island. This would mean covering eighty-five kilometres in five days, much of it over mountainous terrain, and taking in some of the finest scenery on the island.

Under normal circumstances, the walk I had planned would have been entirely achievable, if very challenging, but in view of my badly damaged feet I didn't think I could manage it. For that reason, I knew that I had to cut down both the mileage and the number of days walking. Giving up entirely was an option, but not one that I seriously considered. Having made it this far I was determined to at least make it to the south coast. I would find a shorter route that would at least get me to Las Playitas, and I could see how the feet were making out once I got there. I booked another night in Fustes to give my feet an extra day to recover, and plotted out a route that would still take in the Betancuria Mountains but would get me to

Las Playitas in three days rather than four, and covering forty-five kilometres instead of eighty-five. To make this work, I decided to take a taxi from Fustes to the road north of Triquivijate where I had left it to climb Rosa del Taro on my way to Fustes a few days before. This seemed perfectly reasonable given that my diversion to Fustes had been forced upon me by lack of accommodation in the centre of the island, and to be honest I didn't see much point spending most of a day walking back over the dull eastern plain.

★

The giggling German girls in the apartment next door woke me at 5.30 on Sunday morning. They reminded me of my daughters, no doubt sleeping blissfully in their beds at home. They got me up early, so by 7.30 I was ready to get going, having had a bath, prepared my feet as well as I could, and stuffed everything back into the rucksack. After the luxury of a bed, bath and car, the thought of being back on the road was not an attractive one, and I was quite fearful of what lay ahead. Amongst other things, having thought through the possibilities a number of times I was not sure where I would camp that night. And although my feet felt much better, it seemed only a matter of time before they would be terribly sore again.

After breakfast in the hotel I went out for money and provisions, then sat watching the highlights of an English football match outside a café for a while. I was steeling myself now for the return to the rigours of the walk and rough camping. Then I picked up my rucksack, settled my account and hailed a taxi. The driver showed no surprise when I asked him to take me inland to Triquivijate, neither did he do so when I asked him to drop me at the end of the farm track in the middle of nowhere, about three kilometres north of the village. I paid him and he sped back towards Fustes. The silence enveloped me again. I was back at the point where I had left the road three days before. It was 9.30 in the morning and there was a pleasant breeze. I shouldered my pack and began walking back along the road towards Triquivijate.

★

Not long after I started walking a Barbary Falcon flew over. This bird is like a small version of the Peregrine, being found anywhere in the deserts of North Africa and the Middle East where there are cliffs for it to nest on. It is slightly smaller than the Peregrine, rather paler grey above and creamy rather than white below, although none of this is easy to distinguish as the bird flashes past. Like the Peregrine, it is a supreme flying machine, capable of striking down other birds larger than itself because of its incomparable speed and agility. One of my most memorable birdwatching moments was watching a Barbary Falcon from the top of the spectacular Famara cliffs in northern Lanzarote. One moment the bird was hanging motionlessly in the violent updraft as though it was completely immune to the wind, then it shifted its wing position almost imperceptibly and shot off at a steep angle down the cliff at dizzying speed. Breathtaking! Happily, although the Barbary Falcon has always been quite a rare bird in the Canary Islands, it seems to be doing quite well in Fuerteventura at present.

★

The rucksack seemed almost unbearably heavy, in part because I was carrying nearly three litres of water, but perhaps mainly because I had become unused to carrying it in my days of leisure in Fustes. I stopped for a rest overlooking the plains somewhere to the north of Triquivijate. There were goats bleating on the hill up to my right. Through binoculars I could see that there were two on a ledge of rock halfway down a cliff. It was not at all clear how they got there, or indeed how they were going to get off. The amount of bleating suggested that they weren't too sure either. I scanned the plains below me on the lookout for Houbara Bustards. I had seen one near here the previous year, and there are still good numbers in this area. There was the distant sound of a Sandgrouse flying over the plain, and a near constant passage of Yellow-legged Gulls heading both north and south over the pass, but I saw no bustards. Quite why the gulls pass across the island on such a regular basis is something of a mystery to me.

When I moved on down the road the goats were still stuck on their rock ledge, one bleating and occasionally peering down as if plucking up the courage to jump, the other sitting quietly, either resigned to its fate or unconcerned: it was impossible to know which.

★

Some of the place names in Fuerteventura are clearly not of Spanish origin, and Triquivijate is a good example, as are Tesjuates and Tiscamanita. These place names are of Berber origin. During the sixteenth century, when the supply of pre-Conquest inhabitants had been exhausted and more labourers were required, hundreds or perhaps thousands of Berbers were captured from neighbouring parts of the African coast and forcibly settled in the dry, inhospitable, south-eastern part of Fuerteventura, and the names they gave their settlements are still in use today. Perhaps it is these people who used the wrestling ring, if that is what it was, on Rosa del Taro.

Today, Triquivijate, like so many other places on the island, seems to be something of a boomtown. There are new houses being built everywhere. How things have changed since the 1960s, when John Mercer described Triquivijate as:

'almost abandoned; against the sun, a shack on a hummock, with a single palm, as if cut in silhouette out of gold leaf; then, the gusts of wind sending glowing yellow dust flickering along the distant ridges, ruin after ruin, the amber walls almost visibly slipping and spreading, melting back into the throbbing plain.'

A photograph in his book shows Triquivijate's humble, single-celled church of Ermita San Isidro standing beside a dirt track on the edge of the village, the only building in sight. An old man with hat and stick leads his camel in front of it, delivering water to the houses in the village. Then, as now, the church was behind whitewashed walls with small turrets at intervals, but trees have now been planted round the walls for decorative effect. The humble church's only external feature is a small, functional bell arch stuck rudely on the southern corner at the west end.

There is a road between Triquivijate and Antigua, but I wanted to follow the old mule track that crosses the barren plain about a kilometre further north. In order to do so, I doubled back slightly from the church and found the track heading upwards through a major construction site. On either side for several hundred yards were expensive looking new houses, some still under construction. There were also plenty of barking dogs, so I hurried along, fingering my dog-dazer nervously. By the time I had reached the top of the hill and made it to the safety of the open plain, I was sweating profusely with the effort of hauling the rucksack up the hill in the heat of the day.

The plain now stretched out in front of me offered no shade at all over the four kilometres to Antigua, except for one small, abandoned building, where I thought I might rest for a while. For some reason the door to the house had been blocked with stones, and there was only just enough shade to sit on a less than comfortable rock, looking out over a long-abandoned cactus grove (Box 17).

After a short rest I rejoined the old mule track heading due west, with the Betancuria Mountains on the horizon in front of me. By evening I hoped to be walking amongst them. Crossing the plain in the midday heat, however, proved to be a soul-destroying business, and after half an hour I stopped to take advantage of the little shade I could find behind a low wall. It was a wonderfully peaceful place to sit and rest, the only sounds were the lazy buzzing of flies, the wind whistling gently through the wall, and the distant song of a Lesser Short-toed Lark. Eventually, a lizard came out from amongst the rocks to eye me suspiciously.

I was somewhat refreshed by the time I started walking again, still heading west towards the mountains. There were more larks singing, but otherwise the plain seemed empty and lifeless. But it wasn't: suddenly I was startled from my thoughts as first one, then finally three pairs of Black-bellied Sandgrouse flushed noisily from a scatter of dead weeds at my feet. The nearest of them rose explosively just a few metres to my left, chortling as they went.

Not far beyond the Sandgrouse I was surprised to see human footprints on the track. I realised then that most of my walk was on ground where nobody ever goes any more. Except when I was forced to follow roads, I was either completely off the beaten track, or on tracks that are no longer in everyday use. These mule tracks were once important links between villages, markets and ports, but now they are all but abandoned.

Unfortunately, the first sign of Antigua was an ugly scrap yard. Then came a scatter of houses with the inevitable fierce dogs, all tethered thank goodness, and I reached the welcome shade of trees at the square in the centre of town just after midday. As I did so, a large group of local hikers appeared from the other direction, presumably having walked over the mountains from Betancuria: a popular walk that I would be tackling at least in part later in the day. They mostly seemed to be about thirty-something, and the required equipment was evidently a small day sack, stout boots and a very long, hefty stick. They looked a little like a lost troop of Morris dancers. Exactly what the stick was for I couldn't say, but between them they would have been capable of seeing off a whole pack of fierce dogs. I must admit to feeling rather superior with my heavy pack and lack of weaponry. It was good to know though, that there are local people who enjoy walking. In fact I subsequently learnt that there is a regular programme of walks organised for local people by the island council, a development that I heartily approve of.

The leafy square is the undoubted highlight of Antigua, and a welcome change from the desert landscape that surrounds the town (Plate 29). The lawns are well tended, and there are lines of wispy, conifer-like Casuarinas (natives of Australia, the dry continent), palm trees, and other shade-giving evergreens. On the north side stands the big whitewashed church with its high tower, much more typical of the island's larger churches than the oddity at La Oliva. The tower has large bell openings and a small dome at the top, which makes it look vaguely like a lighthouse.

A number of barrancos converge at Antigua, providing it with storm water from one of the wettest mountain areas on the island. The relative abundance of water is no doubt the reason why Antigua was one of the first settlements on the island, being founded as early as 1485.

★

After cooling down a bit I retreated to the restaurant opposite the church. The walking group had beaten me to it and there were also several large families having Sunday lunches, but I managed to find an unoccupied table with just enough room to stow my rucksack out of the way against the wall. Having done so, I was glad to remove my boots: my feet were already sore after just one morning of walking. The food took a while to arrive because the place was so busy, but I was in no hurry, being happy to while away an hour or so to allow the heat of the day to diminish. I had a large pile of whitebait and plenty of fresh bread, all washed down with half a litre of ice-cold sparkling water. It was simple fare, but just right.

Typically enough for a Spanish family restaurant, the noise level in the place was incredible. In one corner there was a large television, blaring out an American sitcom in dubbed Spanish, of which nobody took any notice. Its only impact was in requiring everyone to shout at each other. After an hour or so I gave up and retreated to the peace of the square.

I sat reading on a stone bench in the shade of a tree, to the heat-deadened cooing of two Collared Doves somewhere on the other side of the square. Above me a big butterfly glided amongst the trees, a Monarch (or Milkweed), the largest of all European butterflies, with a wing-span of around 10cm. In fact, it is not really a European butterfly at all, but a native of North America. It was first noted in the Canary Islands in 1880, having either made its way naturally across the Atlantic during an autumn storm, or perhaps arriving on a ship. Since then it has become established on all the islands, and is found in parks and ornamental gardens. It is easily identified from most other butterflies on the island by its size and gliding flight, but a closer look is required to separate it from the related Plain Tiger (Plate 14). This butterfly, which is only slightly smaller than the Monarch, is African in origin, and native to the islands. The Plain Tiger is found in similar places, though also sparingly around cultivated land throughout the island, popping up in different places in different years. Both butterflies are essentially orange and black with white spots, but the Monarch has thick black marks along the veins in the wings, whereas the Plain Tiger does not, and appears more uniformly orange, hence its name.

★

At 2.30 the newsagent on the square closed. The young woman who had been working there wheeled in the postcard racks, her work done for the day. Another half hour passed, and then a truck heavily laden with bleating sheep and goats rattled through the square. There was a slight cooling and at the same time two or three Blackcaps began to sing from the trees. The character of the square, and of that Sunday afternoon, had changed. It was time to start walking again.

Box 17 Cacti and exotic succulents

The cactus family is native to the Americas, but no less than seven species have been introduced to Fuerteventura and several are now well established in the wild. Originally they were planted to provide impenetrable boundaries to gardens, and for their edible 'prickly pear' fruits. They are well suited to Fuerteventura's climate and produce fruit readily. Although the fruits are protected with spines they can be eaten, once these have been removed, and were probably a valuable supplement to the poor diet in past centuries. In the nineteenth century, cacti became a source of income for the island because of the red dye that was obtained from the cochineal insects that live on them. Large plantations of cactus were created in some areas, notably around La Oliva and in the Antigua/Betancuria region, and many of them still exist. The insects are still there, forming whitish mealy clumps on the leaves, but the dye which was made from them has now largely been superseded by synthetic dyes, and they are no longer of commercial importance. Hence

---

the cactus plantations are now abandoned and overgrown, but since there is nothing to grow in place of them, many have simply been left there rather than cleared away. This is good news for birds such as Spectacled Warblers and Southern Grey Shrikes, for which the overgrown, scrubby plantations provide ideal habitat.

Over the centuries two species of cacti have made themselves thoroughly at home in Fuerteventura. The larger, greyer plant is *Opuntia ficus-indica,* the true prickly pear cactus, which grows wild in *malpais* and even on cliffs, its seeds dispersed from the plantations by birds. The other species is *Opuntia dillenii*, a somewhat yellower and usually smaller plant that can be found growing in an even wider range of places.

In recent years, yet another use has been made of cacti: the creation of ornamental cactus gardens, either on a small scale in private gardens or on a larger scale as a tourist attraction. One fine example of such a garden is just north of Antigua, by the restored windmill, where there are many varieties of cacti and other succulents, and some very large specimens. Particularly impressive are the specimens of the Triangular Spurge *Euphorbia trigona*, an Asiatic species with tree-like masses of organ-pipe stems.

Sixteen

The old capital

From Antigua my route continued west along ancient paths towards the Betancuria Mountains. Just west of the square in Antigua there was a lovely view over a barranco with cactus groves, agaves and palm trees. The barrancos provide a relatively plentiful supply of water for irrigation and there are healthy-looking fields of potatoes and the showy red flowers of Hottentot Fig growing from walls. Amongst the cacti in the barranco is a succulent with discs of leathery leaves at the ends of its bare stems. Sometimes it has spikes of pale yellow flowers. This is *Aeonium balsamiferum*, the only Aeonium found on the island. All known colonies of it in Fuerteventura are apparently near settlements, and given that its sap was used in the past to preserve fishing lines, it seems probable that it was introduced for this purpose. As far as is known, it is native only to Lanzarote. There are no other members of this genus in Fuerteventura, which seems remarkable given that there are no less than thirty-five species of Aeonium in the Canary Islands as a whole. The reason for this paucity of species, of course, is the dry climate and lack of high mountains.

On the edge of the village a Southern Grey Shrike gave its clear two note call from the depths of a pomegranate bush close to the road, before flying out onto the top of a metal wind pump to watch me pass. I could hear its young squawking from somewhere in the dense interior of the bush. Then I passed a new plantation of the latest crop to become important on the island: Aloe vera. To reduce their exposure to the drying winds, the Aloes are planted behind windbreaks, in the same way that tomatoes are. Indeed, they now seem to be replacing tomatoes as the main commercial crop. They are cheaper to grow as they do not need to be watered.

After a while, the track began to rise steeply into the hills. Just where it did so there was a sheltered valley on the left, in which lies the farmstead of El Cortijo Antiguo. The farm buildings and groves are described in detail in Mercer's book, and he considers that it may have been one of the first farms to be settled after the conquest, which is certainly plausible. The wealth of fruit trees and shrubs that still grow here, and the relative lushness of the crops is certainly testament to the value of the site. It is a veritable oasis, with trees big enough to give deep shade, and I was sorely tempted to pitch my tent there and then and rest for the afternoon, cocooned in a leafy paradise. As I contemplated this possibility, a vehicle came up the track from the town and a man began working his crops. I thought of asking his permission to camp, but decided against doing so. Idyllic as it would have been, there were still several hours of daylight left, and as I hadn't covered much ground I decided to head on into the mountains.

With a heavy heart, I carried on up the steep incline, more conscious than ever of my solitude and the barren landscape. I was suddenly envious of the man cultivating his crops in such a lovely spot. I couldn't help comparing it to our unromantic allotment at home in Cambridgeshire. I doubted that he would want to swap!

At first there was nothing but the heat and the sheer physical effort of hauling my pack up the steep slope, but as I climbed higher the view over Antigua and the plains beyond opened out below me, and my spirits began to recover. From here there were views over much of the ground I had covered on the walk so far. Beyond Antigua I could see the peak of Rosa del Taro and the long ridge I had traversed on my way to Fustes. Far to the north was the triangular mass of Muda that I had climbed on the third day. The difficulties and achievements of the walk were spread out before me, but it was the future that played on my mind. I had no idea where I would camp that night, and I could not know what trials lay beyond that. There was nothing for it but to have trust and to overcome any setbacks as best as I could. I would concentrate on now, leave the future to look after itself, and enjoy the moment.

It was at least comforting to be on a path I had walked before: my wife and I walked from the hotel in Antigua to Betancuria and back the year before. As I climbed higher, the bare rocks began to give way to a flower-covered hillside. At first there were just a few flowers here and there: in particular the showy, deep pink Mallow-leaved Bindweed *Convolvulus althaeoides* up to 50mm across, a Mediterranean plant that is common in the mountains. Beside the track the large white flower spikes of the Common Asphodel *Asphodelus ramosus* and, clinging to a rocky slope, the white felted leaves and pale yellow flowers of *Andryala glandulosa,* locally known by the appropriate name *ropa vieja* or 'old clothes'. Here also was my first Painted Lady butterfly of the walk and lots of Green-striped Whites flitting over the flowers. By the time I had climbed above 500m, the heat was rapidly draining from the day and a cool breeze was blowing. Now there were masses of flowers: in places there was almost a carpet of the little pale yellow flowered mignonette *Reseda lancerotae*, each flower glowing in the light of the low sun.

Finally, I reached the top of the hill, and could see Betancuria, the old capital of the island, nestling in the valley below. I headed north along the ridge, keeping Betancuria to my left. I had decided on an unorthodox campsite. The land here was too steep and rocky to camp on, and there would be no benefit in heading down into Betancuria. Instead, it had occurred to me that the tourist viewpoint and restaurant on the top of Morro de Velosa would probably be unoccupied outside opening hours. I knew there was a covered walkway beneath the restaurant, which would be sheltered from the elements. This was virtually the only time on the whole walk that I found myself heading north, and the wisdom of walking north-south was quickly confirmed, as I now found myself battling into a stiff breeze.

In less than an hour, I reached the buildings on Morro de Velosa. I was in luck. In common with most of the museums and tourist attractions on the island, it is closed on Sunday and Monday. On another day it would just have been closing at this time, and there would have been an awkward wait before I could settle in for the evening. As it was, the place was deserted. Having established this, I set about making myself at home. The eastern end of the walkway was the most sheltered, and from there I had easy access to the outer wall on the end of the building from which I could sit and admire the magnificent views, so I left my rucksack under the shelter and took my provisions out onto the wall.

The viewpoint is at around 670m above sea level, one of the highest points on the island, and the views to the north and east are spectacular (Plate 16). It is also one of the windiest spots, but thankfully I managed to find a place where the wind was not too fierce, and having put my fleece on I was reasonably warm. Unfortunately, I then discovered that I had packed my provisions badly. The bananas and pear had disintegrated into a semi-liquid sludge that had coated the inside of my little day-sack. Fortunately, most of the rest of my provisions were in better shape, and I was able to find enough edible remains for a modest snack.

There were a surprising number of birds. An African Blue Tit was taking food into a nest in the wall just below the restaurant, a Barbary Partridge was giving its steam train-like call from rocks just below me, and a Corn Bunting was singing just behind the restaurant.

As the light began to fade, the street lights came on far below in Antigua and Llanos de la Concepcíon, mirroring the stars that were beginning to appear above. The moon, which had been new at the start of the walk, was now over half-full and almost directly overhead. It was a wonderfully peaceful spot to watch the night descending on the island, though the mountain air was becoming distinctly cold.

It was almost completely dark when I returned to the shelter of the passage under the restaurant to decide where to sleep. I chose a spot at the bottom of a ramp which was fairly sheltered but where I could see the moon and stars above me. I dozed with the sound of the wind howling around me, snugly wrapped in my sleeping bag with the moonlight on my face. At some point in the night, the wind shifted direction and I woke up cold, so I retreated to the covered area, but in a place where I could still see the moon.

Next morning I had a rudimentary breakfast of dried fruit and water, back at the spot where I had watched nightfall a few hours earlier. Now the sun was rising above a thin layer of clouds over the sea towards Africa. It was undoubtedly beautiful, but I felt lonely. Still, I was looking forward to the walk down to Betancuria, where I hoped I would find coffee and something more substantial to eat. There is a road from the viewpoint to Betancuria, but it is busy with traffic and would not be a pleasant road to walk along. I therefore set off along the old mule track on the eastern side of the valley, which must once have been the main route linking Betancuria to the northern part of the island.

★

The old track sloped gently downwards between dark green shrubs, from the depths of which Sardinian Warblers scolded and rattled, just as they would have done as mules made their way over the pass in days gone by. But the other sound that accompanied my walk to Betancuria is a new one in these parts: the acclaimed song of the Canary, with its incomparable trills and fluty whistles. This bird is native to the islands further west, but has been deliberately introduced to Betancuria over recent decades and has established itself amongst the lightly forested hills.

The introduction of animals and plants to places where they are not native is generally frowned upon, and there are certainly plenty of examples where introduced animals have caused serious problems. For me, though, the introduction of the Canary to Fuerteventura is a rare example of a good introduction. The island has a relatively poor avifauna, and almost all the native species are birds of the open desert. This being the case, the newly forested areas are mainly devoid of birds, and the Canary seems almost to be a necessary addition. It has certainly made itself at home in this mountain enclave, and its glorious song definitely adds something to the countryside hereabouts. And perhaps the Canary lived amongst the native olives that grew here before they were cleared by man, in which case it has simply been brought back home.

The walk down the mule track was as delightful as I had hoped. The sides of the track were dotted with flowers, and as Betancuria came into view the sound of church bells mingled with the songs of the Canaries. Nearer to the town, majestic agaves planted alongside the track added a sculptural touch to the approach (Plate 17), and reassuring bird sounds drifted up towards me: the purring of Turtle Doves, the contented chirping of Spanish Sparrows, and the insistent cooing of Collared Doves.

★

Betancuria was the first capital of the island, and is named after Juan de Bethencourt, the Norman nobleman who played an important part in the initial conquest and settlement of Fuerteventura by Europeans in 1404, and who established Betancuria as his base on the island. It is about 400m above sea level, surrounded by mountains and therefore not so easy for sea-going marauders to plunder, an important consideration at the time.

On the northern edge of Betancuria is the ruined Franciscan convent of San Buenaventura and a small associated chapel, the Hermitage of San Diego de Alcalá, which is still intact. These buildings are easy enough to reach from the road, but not from the mule track. To reach them I left the track just north of the village and crossed the barranco, which proved to be more challenging than I had expected. The barranco here is choked with Giant Reed and tall shrubs, and I found myself having to force a way through a near impenetrable tangle of vegetation, something I hadn't expected to have to do at any point on the walk, and I almost fell over a small agave that I had not noticed. Had I done so, the plant's spiky leaves might have given me a nasty wound.

Around the hermitage and convent are informal gardens dedicated to the 'Conquistador García de Herrera'. A small convent was first established here soon after Betancuria was founded in 1414. It was enlarged during the time when Diego García de Herrera was Lord of the Canary Islands in the latter part of the fifteenth century. He arrived in Fuerteventura in 1454 with the friar Diego de Alcalá, who was subsequently canonised, hence the name of the hermitage. From Betancuria, García de Herrera oversaw raids on the islands to the west, and in 1476 he established a trading post on the African coast opposite the island. The primary reason for doing so was to capture slaves. However, in 1524 the African coastal settlement had to be abandoned, due to a combination of disease and local hostility. Thereafter, the position was rather reversed, and the Berbers took to raiding Fuerteventura. In spite of Betancuria's relatively safe location, it was sacked by the Berbers in 1593, when Bethencourt's original church was destroyed. This, however, is not the reason for the ruined state of the convent. This came about due to the outlawing of the Franciscans in Spain in 1820. At some point thereafter, the convent evidently became disused and the building deteriorated. Today it is just a partially restored shell. Nothing remains of the roof, but the outer walls are in good repair and its massive gothic arches are intact.

★

The village itself is dominated by the church of Santa Maria, which is on the north side of the village square (Plate 30). The square forms the nucleus of the old capital, which explains why there are so many fine old two-storey stone buildings with balconies. Amongst them is the restored Casa Santa Maria, with its café and pretty garden which is now open to the public. To me, the village has a distinctly more European feeling than any of the others on the island. Most are dominated by single storey buildings and would not look out of place in Morocco.

My first priority was to find a café that was open. The bars and restaurants around the square were all closed – Betancuria caters mainly for tourist coaches – but fortunately the one on the street above the church was open, although there were no other customers at its rustic bar at this time of day. The tourists had yet to arrive. The building appeared to have changed little over recent decades and was in keeping with Betancuria's quaint aura.

Unfortunately, my feet were already very sore, and I was in no rush to start walking again. I passed an hour or so enjoying good coffee and reading recent copies of the local paper *Canarias 7*. My attention was drawn first to the football pages. I was impressed by the high standing of some of Fuerteventura's teams, reflected both by recent results and their positions in the national leagues. Football clearly matters here. There was also an article about the new ferry service between Puerto del Rosario and Tarfaya on the African coast. It has always been something of a surprise that there is so little contact between the island and Africa, given the proximity: the island is only around 100 kilometres off the coast. There was a good deal more contact during the fifteenth and sixteenth centuries then there has been in recent decades! The ship *Assalama* began sailing between Fuerteventura and Tarfaya towards the end of 2007, but the conveyance of goods had not been approved in March 2008 and there were fears that the service would be suspended. In fact, the ferry link ended in rather more dramatic circumstances, when the ship hit a rock off Tarfaya on 2nd May 2008, and the 113 passengers on board had to be rescued.

Having read the papers, I wandered back down to the square to sketch the church. After its demise at the hands of the Berbers in 1593, the original church, then the island's cathedral, was left in ruins for a few decades before being rebuilt some time before the middle of the seventeenth century. Why it was not immediately rebuilt does not seem to have been recorded, but presumably there were fears that the Berbers would return, and in any case the convent church could have been used in the meantime.

The gleaming, whitewashed building that stands here today is therefore mainly seventeenth century. Of all the churches on the island, it is certainly the grandest and has an incomparable setting, perched as it is above a verdant barranco, amongst pleasant old buildings, and with a rounded mountain behind. It still has the presence of the cathedral it once was. It is sturdily built with small windows placed high in the walls. It seems as though the threat of attack was still real when it was re-built. And yet the proportions of the building are right: the tower is just large enough to counterbalance the sturdy nave, whilst the old whitewashed houses along the west side of the square are of similar height to the nave and add to the overall appearance.

In common with many of the older buildings in Betancuria, the church's larger corner stones are left bare to create a simple but effective pattern against the otherwise completely whitewashed walls. A curious feature of the tower is that the stones in the lower part below the first window are darker, and appear to be of lava, whereas the stones above, as well as on all the other corners of the building, are of the more usual reddish-coloured rock. Perhaps these darker rocks remain from the original fifteenth century church. The tower has a roof covered with red, yellow and cream tiles, from the centre of which rises a structure that looks as though it is made of copper with wooden slats, topped with a fine weather vane, complete with cockerel.

The most striking feature on the exterior of the church is the wonderful Baroque doorway. The pilasters on either side are carved with plants that have elegant leaves and sunflower-like flowers level with the top of the door. Above the flowers are two capitals of different design separated by unusual diamond-shaped carvings. Above the door is a triangular pediment with a coat of arms in the centre, above which is a small window. The wooden door itself has carved panels.

Looking closely at this doorway, some of the elements of the bizarre decoration at nearby Pájara Church can be seen: sunflowers and diamonds form part of the doorway decoration, and the carved panels have a form not unlike the curious carved designs above the Pájara doorway. This is not to suggest that the carvings at Pájara are entirely inspired by the Betancuria doorway, but there are certainly elements that might have been part of the inspiration for Pájara's immensely imaginative work. Given that the two churches are only a few miles apart, it would hardly be surprising if this were so.

★

At one time there was a good restaurant on the main street, with a covered veranda that had delightful views across the heavily vegetated barranco to the church. During my period of study on the island, a meal at this restaurant had been the ultimate treat. On one particular occasion, when my school pal Nick Parker and his wife Sheila were staying with me, we had somehow ended up having rather too much to drink here. This was in no small part down to the very friendly patron, who insisted on giving us coffee laced with copious quantities of cognac, and this after I, for one, had already had more than enough red wine. Our host recognised me as the mad cyclist from La Oliva – it seems my fame had spread throughout the island!

By the time we had finished, none of us was in a fit state to drive home, so we found a grassy verge for an afternoon nap. After quite a while I was vaguely aware that Sheila, who had sensibly drunk the least, had gone for a walk, and came back much refreshed. She had walked to the top of the opposite hill and back, and the exercise had cleared her head. Nick then did the same, and after a while he also came back, confirming that the walk had beneficial effects, and encouraging me to do the same. My head told me that I was in no fit state to walk anywhere, least of all up a hill, but I dutifully hauled myself to my feet and set off up the opposite slope. At some point I remember coming to a crest and deciding it would be a good spot to sit down for a bit before returning. From below it must have looked as though I had simply fallen, because the next thing I knew a worried Nick had hauled himself back up the hill and was asking me whether I was all right. He drove us back to La Oliva, where I clambered into bed and shivered uncontrollably for some hours before falling asleep and recovering. On reflection, perhaps I was suffering as much from sunstroke due to falling asleep in the sun as I was from the effects of alcohol.

★

The friendly patron of that old restaurant also used to be the keeper of the archaeological museum over the road. You had to get him to open the museum specially, as indeed was also the case with the church, but it is now open on a similar basis to the other museums around the island. It contains an interesting display about the pre-Conquest inhabitants, together with examples of pottery and various other artefacts found in their dwellings. The first evidence of humans on the island are the 5,000 year old remains of a goat, which must have been brought to the island by the first settlers. It is not clear whether the island was occupied continuously from that time. The most interesting set of artefacts are from a cave in Montaña Arena at La Oliva, where six crudely sculpted figures made from bones were found. They are two to three inches high and hardly recognisable as representations of the human form. If these are the high point of artistic achievement in the pre-Conquest people, then clearly their society was very primitive.

Today the museum was closed, so I headed down the road in search of provisions. To my relief, almost the last building by the roadside had a shop and a small café, both of which were open. After buying food and water I had tea on the enclosed terrace, which had a wonderful view. In the foreground was the reddish earth of a bare field and a barranco choked with giant reed and trees. Beyond, and framed between the foliage of two big palm trees, the whitewashed houses of the old town and the church tower dominating all. But now it was time to leave the old capital behind and head south once again.

Seventeen

Pilgrimage

Just south of Betancuria, I passed an old woman wearing the traditional straw hat and walking purposefully towards the village with the aid of a stick. She was one of the very few people I saw walking in the countryside on the whole journey. Shortly after passing her, I turned left onto a track and followed it across the barren hillside towards the Medio Ambiente buildings, where a range of native Canary Islands plants is being grown. Medio Ambiente is the government body responsible for nature conservation. Just beyond these buildings the island's only 'forest' came into view. The Castillo de Lara forest is a sparsely timbered area covering about a square kilometre of steep hillsides. In any part of mainland Europe this would be passed off as a wasteland, but here in Fuerteventura, the sight of so many trees in one place is a shocking visual spectacle. The area was planted with a variety of non-native pines and acacia trees in the late 1940s to see whether trees could be grown commercially here, in this wettest part of the island. The trees are mainly Aleppo Pines and Monterey Pines, with some Canary Island Pines. The last species is native to the mountains of the islands west of Fuerteventura, where they form extensive forests of stately trees above 1,000m. It is an unusual pine tree in that its needles are in clusters of three, and its closest relative grows in the Himalayas.

Sixty years after they were planted at Castillo de Lara it is clear that the pines will never provide a commercially viable crop. Many have died, and those that are still alive have sparse foliage that, at best, provides a little dappled shade. The conditions are just too harsh for them here, so they grow very slowly and do not regenerate. More happily, in late 2007 hundreds of native trees and shrubs were planted on the high ground above the forest and in other high land around the island. Although this is a small beginning, these trees will do much better than the pines, and it is hoped that they will eventually re-create a semi-natural forest that will be self-regenerating.

Carrying on down the track towards the forest picnic area, I noticed a few patches of one of my favourite Fuerteventuran plants, *Caralluma burchardii,* a succulent member of the milkweed family. It has stout square stems about an inch across and a few inches high, sometimes greenish but more often dull grey. It spreads by means of underground stolons, so tends to grows in clusters of ten or twenty, looking for all the world like clumps of sculpted rock. Somewhat infrequently it bears rather striking clusters of star-shaped purplish flowers with orange-yellow centres (Plate 23). Typically for a member of this family, it then produces a pair of pods that can be four inches long, looking remarkably like horns.

Although it also occurs in adjacent parts of Morocco, within the Canary Islands *Caralluma burchardii* is found only in the eastern islands, and is much more abundant on Fuerteventura than Lanzarote. It is considered to be in some danger due to habitat loss, and is listed on Annex II of the EU Habitats Directive. This means that the Spanish Government is required to identify protected areas for it, although sadly this has not stopped some colonies from being bulldozed out of existence. The Betancuria mountains are a stronghold for it, and it is also common on the *malpais* near La Oliva.

Another plant that was growing beside the track was a fine shrub more than a metre tall and perhaps twice as wide, and covered in spikes of white flowers. This was *Echium decaisnei,* probably not the subspecies native to the Jandía mountains and northern Lanzarote, but the introduced form from Gran Canaria.

★

As I approached the picnic area, a scrawny terrier came flying towards me growling in a blood-curdling way that seemed to confirm an intention to bite. I fingered my dog-dazer nervously, but the Medio Ambiente rangers who were working there, and to which it seemed to belong, shouted frantically at it and it stopped just short. I chose a picnic table which seemed to have reasonable shade. My intention was to rest there through the middle of the day before heading up through the forest and out onto the high ground beyond.

No sooner had I settled at the table than the terrier came over to see me, encouraged perhaps by the sight of my lunch. Having got over its initial excitement it now decided to befriend me, and its ambition for the early afternoon seemed to be the same as mine. Once it realised I wasn't going to feed it, it sprawled in the shade under the table and dozed.

Lunch was a hot affair. The shade provided by the wretched pines was so sparse that I had to keep moving around the table in order to get some relief from the hot sun. Despite the heat, Ravens were noisy. Quite what they were about I don't know, but their gruff conversations reverberated through the forest all through the afternoon, and now and then came the sound of their wings beating the air as they chased each other through the trees.

Eventually, after much calling from the Medio Ambiente rangers, the dog reluctantly got to its feet, stretched, yawned and trotted off towards them. I was sad to see it go, as I knew it would be my only companion for the rest of that day and probably for most of the next.

★

Some time later a noisy family took up residence at a table just below me, and I decided it was time to move on. Before darkness fell, I needed to cover about seven kilometres of rough ground, rising over some of the highest points on the island, and I was keen to get going. The first task was to climb up through the forest to the ridge above me. The path was really steep, and within minutes I was sweating profusely in the hot sun. I hauled myself up the slope until I came to the forest edge, where I found a shaded rock to sit on. There was a splendid view over the forest, with Canaries singing all around me and African Blue Tits chattering in the valley below. Then I set off again, negotiating my way through a hole in the forest's boundary fence to attain the ridge. Three Buzzards glided past going north.

It was a pleasant surprise to find a good track heading south along the ridge, but surprise would be too mild a word to describe my feelings when, a few hundred metres further along, I came upon a well-constructed viewing platform! This was not something I expected to find on a remote dirt track, and the reason for its construction was difficult to understand. I took off my pack and stopped to admire the view over Antigua and the central plain and to watch yet another Buzzard drifting past at close range. When I picked up my pack I noticed that it was covered in red spider mites. In the process of brushing these off I discovered that my rolled up sleeping mat, attached to the underside of the pack with shoe-laces, was about to fall off. Thanks to the mites, disaster was averted.

The first peak I came to on the ridge was Morro Janana (674m), where there were no less than fourteen assorted radio masts, some quite new but others probably redundant. The track ends here, and beyond there was just a vague path, hardly discernible at all in places as it wound between the stones. Directly ahead, four kilometres along the ridge, was my next goal, the peak of Gran Montaña, at 708m one of the highest peaks on the island.

Unlike the thrilling knife-edge ridges on the eastern side of the island, this was much less threatening, and the slopes quite manageable. The mountains around Betancuria are older, of different geological origin, and generally more rounded. It would be fairly easy walking most of the way, and I could see from the map that the ridge never dropped below 550m. Nevertheless, I was very conscious that I was heading into wild, remote country, and I couldn't help feeling apprehensive.

The only significant peak between Morro Janana and Gran Montaña is Morro de Tabagoste, which is unremarkable in all respects except one. Whilst the east-facing slope has the same barren appearance as its surroundings, its west face is striped with neatly planted rows of dense, dull green acacia bushes, for all the world like cross-hatching on a map made real by a giant artist. Just beyond Morro de Tabagoste I sat down for a while to admire the views over the Río Palmas valley. With its stately palm trees, this is the loveliest valley on the island. The church in the valley houses the patron saint of the island, and there is a pilgrimage to it every September (Box 18).

Two Egyptian Vultures drifted past and high above, the air was busy with swifts. Somewhere in the acacias below another Canary was singing.

★

It was now late in the afternoon and the sun had disappeared behind clouds so it was much cooler. Perfect walking conditions, in fact, for the last two hours of the day. The ridge rose fairly gently now towards Gran Montaña, providing a relatively easy approach to the summit. At about 600m there was noticeably more vegetation, the result no doubt of slightly higher rainfall. Most conspicuous was *Asteriscus sericeus,* a handsome silvery-leaved shrub with yellow daisy-like flowers up to 3.5cm across (Plate 16). This plant is endemic to Fuerteventura, and is common in the Betancuria mountains. Other plants included *Micromeria varia,* a thyme-like plant with little mauve flowers that grows amongst the rocks throughout the upland areas of the island, and *Rutheopsis herbanica,* an endemic umbellifer with characteristic pinnate leaves.

About 50m below the summit I reached the watershed, with the peak of Gran Montaña itself rising steeply to my right, and views over the southern flank of the mountain in front of me. I stopped to assess the two possible ways down. From the peak itself a ridge ran due south, whereas to my left a second ridge headed to the south-east, more or less directly towards the village of Tiscamanita, through which I would be passing tomorrow. On the map the ridge due south of Gran Montaña had seemed the more likely route, but it now looked as though the south-easterly ridge was the easier of the two, since it lacked the steeper sections that I could see on the southern ridge. Having made this decision, the summit of Gran Montaña was now a diversion from my route, so I took off my pack and scaled the last steep incline un-laden. How gloriously easy it was without the pack, and in five minutes I was at the top.

The view from the peak was not particularly spectacular: the slopes before me were relatively smooth, and it was quite hazy. Below were the villages of Toto and Pájara, and in the far distance I could for the first time just make out the south coast and the white buildings of Las Playitas, where this section of the walk would end some time tomorrow afternoon. The thought of a bed and a few days rest was certainly an inviting one. To the south east was the village of Tiscamanita and far beyond it, through the binoculars, was the track across the plain that I would take the following day. Looking back the way I had come, the silhouette of the building where I had slept the night before marked the opposite end of the ridge seven kilometres to the north, and beyond, in the hazy distance, was my last sight of the triangular peak of Muda.

After I had enjoyed the view, I started back down the slope, retrieved my pack and headed down the ridge towards Tiscamanita. After a few hundred metres I glanced back at Gran Montaña only to see that there were, in fact, two peaks! Had I climbed the higher peak or not? Surely I would have noticed if there had been a higher peak in front of me when I was at the top, but from the south, perspective made it look as though the higher peak was in fact the more southerly one. Looking again at my trusty map, there was definitely only one peak marked, so I had no way of knowing whether I had got to the top of Gran Montaña after all. Fortunately, when I had time to unearth the old military map from the recesses of my pack, it showed a second, slightly lower top to the south of the main peak. I had indeed climbed the higher peak.

My decision to descend via the south-east ridge proved to be an excellent one. Wonder of wonders, there was actually a path of sorts to follow. It was certainly not a well-trodden path, more like a well-used goat track, but it avoided the more difficult rocky parts of the ridge by taking a slightly lower line on the leeward side. This was the first and only time I found a footpath in such a remote area.

At first I couldn't understand why it was there, but then I realised what it was. The path led directly from Tiscamanita, north-west over the mountains to the pilgrim church at Vega de Río Palmas. Along this path, the distance between the two villages is only about seven kilometres, whereas by road it would be twenty. Admittedly, walkers following this path would have to climb over the ridge, an ascent of 450m from the village, but this would simply make the journey more of a pilgrimage. And walking the narrow mountain road with its heavy fiesta traffic would be very unpleasant and quite dangerous.

The path skirted round the edge of an old cactus plantation with sentinel agave plants marking the way, and it zigzagged to keep the incline manageable (Plate 18). Before me was the utterly barren slope of Carbón, the southernmost outlier of the Betancuria mountains, and the whole of the central plain spread out below. The views were wonderful, the going was unexpectedly easy, and I was in heaven. At that moment, this was *my* pilgrim path, a pilgrimage of a different kind to the one it had been made for, guiding me gently downwards on my personal journey. I found myself first humming the only song I knew that contained the word 'pilgrim', the hymn 'He who would valiant be' and then singing it at the top of my voice. It was a rare moment of pure elation.

It was a hymn we sang often at school, and I could more or less remember the words of the first verse, which are as follows:

He who would valiant be 'gainst all disaster,
Let him in constancy, follow the Master,
There's no discouragement shall make him once relent,
His first avowed intent, to be a pilgrim.

The words are from *Pilgrim's Progress*, written by John Bunyan in 1684. My paternal grandmother always insisted that her family was descended from John Bunyan, though I have seen no evidence to confirm this. If indeed he was an ancestor, how fitting that his words should come to me in this moment.

★

The light was fading fast by the time I neared the bottom of the mountain, and I needed to find somewhere to camp. Down to my left, an earthen dam had been constructed across the barranco, and the silt deposited behind it formed a flat area that looked perfect. I made my way down the steep slope for a closer look. It was sheltered and the deep, alluvial soil was ideal for pitching the tent. There was quite a lot of annual vegetation, including the poisonous yellow orbs of gourds *Citrulus colocynthus*, the size of tennis balls. The ground was heavily trampled by goats, and I could hear them not far away on the outskirts of Tiscamanita. I guessed they probably came through here each morning on their way to graze in the hills. It would be as well to rise very early to be out of the way before they appeared, and I wanted to make an early start anyway, so this wouldn't be a problem. As the last light of the day ebbed away I lay snug inside the tent, tending my blistered feet as best I could with soapy water and creams, though I knew there was nothing I could do to prevent the next day being painful. It had been a relatively easy day, and I had covered just over twelve kilometres. I would have to cover almost twice the distance the next day on rapidly deteriorating feet. I would rise as early as possible to make the most of the cooler part of the day. At least I had a bed and what promised to be a good hotel to look forward to at the end.

After dark there was a symphony of crickets to sooth me, and a Stone-curlew called, some way off to start with, but then very close. It was clearly feeding on the flat ground around the tent. Despite these reassuring sounds, whether it was because the tent was flooded with moonlight or not, I don't know, but I found it hard to sleep. I lay awake much of the night worrying about nothing in particular. How fleeting elation is, and how readily the mind lingers over uncertainties.

Box 18 *The Río Palmas Valley and Our Lady of the Rock*

The Río Palmas Valley is one of the most fertile parts of the island on account of the relative abundance of water. The surrounding mountains benefit from relatively high rainfall, and much of it is channelled into the valley by the topography. No surprise, then, that it has some of the finest palm trees on the island. These are both the native Canary Island Palm *Phoenix canariensis* and the introduced Date Palm *Phoenix dactylifera*. It appears that the date palms were already here when European settlers arrived, so must have been introduced by the pre-Conquest islanders from their African homes.

The Las Peñitas reservoir was constructed in the valley in the 1950s to retain storm water. Although it is now full of silt and is virtually useless as a reservoir, there is a delightful walk down the tamarisk-lined barranco from Vega de Río Palmas. There is usually at least a trickle of water in the barranco, and although the reservoir is normally completely dry these days, there is still a dense growth of tamarisk and reed. Below the dam a track leads to the tiny Hermitage of Nuestra Señora de la Peña (Our Lady of the Rock), nestling on the side of a steep ravine and overlooking pools of water. The simple whitewashed chapel was built on the site where the statue of Nuestra Señora de la Peña was found in 1443.

Leaving aside its religious significance, the statue in question is of genuine interest. It is made of alabaster, and was carved in France in the late fourteenth century. It was brought to the island by Bethencourt in 1402, when he and others were trying to establish a settlement. The chronicles of the time tell us that they left in something of a hurry, and many of their belongings were left behind, including the statue, which it seems was buried at the narrow gorge where it was subsequently found. In view of its significance and value, it seems to me that it would be surprising if the fleeing monks had not made a careful note of where they had hidden it. If so, the returning monks presumably knew where to find it, but legend has it that it was found miraculously after they had been led to the spot by heavenly lights.

Nuestra Señora de la Peña became the patron saint of Fuerteventura, and in the 1880s a tradition of pilgrimage to the site began. At first, the pilgrimage was made on foot, or with the help of a donkey or camel. Today, the statue is in the church in Vega de Río Palmas. On the third Saturday in September, the church hosts the most important religious festival on the island. In 2008, for example, it is estimated that 25,000 people made the pilgrimage. They come from all over the island (and indeed from other islands), mostly arriving in special coaches, but many still on foot. One popular tradition is to walk over the mountains, from Antigua along the track to Betancuria, with much singing and dancing.

EIGHTEEN

A *low point*

It was still dark when I woke from a fitful sleep, but the early risers were already stirring. A Trumpeter Finch called from the cliff above the tent, and the distant yowling of a Peacock was mixed with the more typical crowing of cockerels in the distant village of Tiscamanita. I had a basic breakfast and broke camp as quickly as I could. The street lights were still on in the village below when I regained the track and clouds billowed down the nearby mountain slopes. To the south there were showers over the sea in the distance, with streaks of rain that evaporated in mid-air. The south coast rarely receives meaningful rain.

The streets of Tiscamanita were almost empty, and the few people who were about were probably very surprised to see me, though they hide their curiosity well. To my delight the bar Tio Pepe on the main street was open. My wife and I ate a good traditional meal here one evening when we were staying up the road in Antigua. In fact, the bar was not only open, it was full to bursting and very lively, with men coming and going all the time. The contrast with the quiet streets could not have been greater. It was with some difficulty that I managed to order coffee and find room to sit and drink it. At one point a woman poked her head in to ask something but quickly disappeared again. Perhaps she was inquiring after her husband, as this was clearly not a woman's domain. Tiscamanita is still very much a working village, and the atmosphere reminded me very much of La Oliva thirty years ago.

Unfortunately there were no shops open at this hour, and I learnt that none would be open until nine. This should not have surprised me, but I could not wait that long as I needed to get as far as I could in the cool hours. The bar sold me a litre and a half of water to add to the litre I had left. My only remaining food was some chocolate, dried fruit and a pear, which was not much for a twenty-kilometre walk, but at least there was the promise of a good meal at the end of it. Also, it looked as though there might be a restaurant of sorts at the settlement of Teguital, about three-quarters of the way to La Lajita. With luck it would be open and I could stop there for lunch. Nevertheless, I dared not rely on it in such an isolated spot away from an obvious tourist trade.

There were lots of old stone buildings along the road heading south-east out of the village, most of which were in ruins, though a few were still occupied. There were lots of Hoopoes too, which no doubt found plenty of nest sites amongst the ruins. One immaculately whitewashed house, set back from the road amongst withered pasture and a few cacti, was larger and clearly designed to impress. The single-storey living quarters were half-hidden behind what appeared to be a low outhouse, at either end of which was a rather grand, squat tower topped with pyramidal castellations. Each tower also had a single shuttered window. The left-hand tower was clearly a dovecot, with holes irregularly placed on either of the visible sides, some at least of which were occupied by doves. The overall appearance of this curious building was suggestive of the Casa de los Coroneles in La Oliva, on which it was presumably modelled.

Just to the north was the impressive volcanic cone of Montaña Gairia, perhaps the most perfectly shaped volcano on the island. The old houses straggled on for a kilometre or so, and the last on the right was still in use. The yard was full of goats, and the shy goatherd watched me pass and hoped not to be seen. Beyond that there were a few modern houses that looked completely out of place, and then nothing but desolate plain. I left the road to follow a track towards Casas de Ezquén, seven kilometres to the south-east.

Back in February 1981, when my school friend Nick Parker was staying with me, we had driven down this track in search of bustards and slept in the car. The following morning the car wouldn't start, so we had to walk back into Tiscamanita in search of assistance. We got a lift to Tuineje with a Chilean man who now lived on the island, but there was no mechanic in Tuineje who could help, so we got a taxi back to Tiscamanita and rang the car hire firm. By the time the mechanic arrived, most of the morning had gone. He started the car first time!

★

After about two kilometres, in a particularly dreary bit of the central plain, was a large fenced-off area some distance from the track, where Egyptian Vultures are provided with carrion. Beyond the fence (and locked gates) was a substantial observation hide, which is only accessible if you have permission. The vulture feeding programme was set up with the assistance of EU funding, and has resulted in a significant improvement in the fortunes of the vultures on the island. At one time they were quite scarce, with only about twenty breeding pairs remaining on the island, but thanks to the provision of supplementary food they now seem to be slowly increasing. There were several hanging around here, and I saw more during the rest of the morning.

Further on there was an area of relatively lush pasture off to the left, below the craggy edge of the lava field of Malpais Grande. As its name suggests, this is the largest lava field on the island. On the far side of the pasture there was a ruin and a solitary palm tree, which was full of chirping sparrows: a moment of vitality in an otherwise dead landscape. The 'fields' were fenced off, otherwise, despite my painful feet, I would have made a detour to enjoy being with the sparrows for a few minutes in the shade of the palm tree. But alas, my way was over the bare plain, now heading towards the curiously flat-topped volcanic cone of Caldera de Liria. To its north stood the higher and more usually shaped cone of Caldera de La Laguna.

As I got nearer to Caldera de Liria, the reason for its strange appearance became clear: it was being taken away! High on the cone's flank a digger was eating away at it and a procession of lorries was taking the cinder to a building site somewhere on the coast. The price the island pays for tourism. The reason for singling out this cone for destruction was clear enough: it is a long way from any roads and hidden by the landform so the impact on the landscape is minimal. But which cone will they take away when this one has gone?

Just before the flattened volcanic cone, the track passed over a slight rise and dipped suddenly into a barren little valley, with rather fine views south-east across the plains to the coastal mountains. I had already covered six kilometres since leaving the café in Tiscamanita, so I felt justified in stopping for a light snack. There were no less than five Egyptian Vultures and two Buzzards hanging around the disappearing caldera, otherwise there were no birds here at all.

Further down the track I had a surprise. A few hundred metres ahead of me was a camel. Through the binoculars I could see that it was neither tethered nor hobbled. I hoped it was friendly as I would have to walk right past it. Just in case it wasn't, I picked up a dead tobacco stem about a metre long. In fact, it simply watched me pass without moving at all. It seemed as surprised to see me as I was to see it, and unlike the local people it wasn't embarrassed to stare!

Up until the early part of the twentieth century camels were important to the island's economy. They provided an effective way of distributing goods around the island and were also used to plough the fields after winter rains. By the early 1980s, they had been almost entirely superseded by motorised transport and tractors, but a handful were still used for ploughing. I twice saw a camel and a donkey tethered together and used as a ploughing team, once near Tetir and once at La Oliva. Perhaps the La Oliva animal was the one that destroyed my cardboard box bird hide. Another decade or so and they would probably have disappeared from the island forever, but then came tourism. They are no longer of any use in agriculture, but in one or two places on the island they provide novelty rides. When I brought my daughters here we rode on them at the zoo at La Lajita, and enjoyed the experience very much. The zoo now also has a 'camel sanctuary', where there are many dozens of them.

Whilst the tourist camels are strong and well kept, this poor beast was in a bad shape. Its back legs looked all wrong and I wasn't sure whether it could have run after me even if it had wanted too. Possibly it was maltreated, but I prefer to think it was just very old: the last, perhaps, of the islands working camels, feral now and living as best it could in the desolation. There would be enough vegetation to keep it going, but where does it find water?

★

Not far beyond the camel was the entrance to the cinder quarry, the point at which lorries joined the track, though thankfully they were not too frequent. My feet were already very sore, and it was getting hot. The landscape was more barren than ever. There was nothing for it but to grit my teeth and plod on. The one consolation was that I did at least have more than two litres of water, so there was no risk of getting dehydrated. I was learning!

The track was almost a road now, having been widened and improved for the lorries. After a while it swung left across the lava field of Malpais Grande. The lava here is not as raw as the new lava on Lanzarote, but it is still an inhospitable jumble of sharp rocks with little vegetation. It is almost, but not quite, impassable on foot. One day back in 1979 during the Houbara expedition, Dave Shirt and I were dropped on a track somewhere near here and asked to walk in a straight line west across the plain towards Tuineje. There was no proper road here in those days, just a maze of tracks, and we had been dropped in the wrong place. Our transect went straight across the *malpais*. It was hard to find a way through the boulders and thorny *Lycium* bushes, but with fresh legs and a light pack it was actually quite an interesting experience. I remember it being something of a relief, in fact, after days of walking across plains.

★

Halfway across the *malpais*, there was a dead sheep beside the track, its face bloody and its eye recently pecked out, probably by a Raven. But then, horror of horrors, I saw that the poor beast was not dead:

it was taking shallow, rapid breaths. Flies swarmed round its bloody nose and eye and tormented it terribly. It could just manage to flick its ear from time to time and move its nostrils to stir the flies, but the relief was almost none, because they settled on her again almost immediately. She must have been hit by one of the lorries. I crouched close, talking softly to her, wishing her a swift death. I became tearful as I explained that there was nothing I could do, then walked sadly away down the track. Bizarrely, ten metres or so further on there was a little wooden cross on a rock.

It was late morning by the time I reached the FV2, a very busy road these days forming the main link between the airport and the tourist developments in Jandía. Unfortunately, there was no choice but to walk beside it. There was nothing scenic about this section, just barren land with scattered houses, and almost constant traffic. It was simply a matter of covering the three kilometres to the little settlement at Teguital as quickly as possible.

When I finally arrived at Teguital, my fears were confirmed. The bar was no longer in use. I was at the limit of my energy and I needed somewhere to rest for an hour before the last push to the south coast. Over the road from the bar there was a stone bus shelter that would at least provide shade, although no respite from the roaring traffic.

Lunch, if it could be called that, consisted of a badly bruised and mushy pear, a few squares of chocolate and a handful of dried fruit. Then I sat reading the wheelbarrow book to try to lift my spirits, but fate was against me. After just a few minutes a police car pulled up. For a moment I wondered whether they were interested in me, and I steeled myself for a discussion about wild camping. But no, they were traffic police and they had come to set up a lorry check-point. The heavy vehicles parked a few metres from me whilst forms were examined and questions asked. Misfortune was turning into farce. Roaring traffic I was prepared to put up with, but rest and recuperation were now impossible. Despite my weariness I had no choice but to carry on south through the hottest part of the day. The journey had reached a low point.

Although my map showed a track heading south, the only one I could find was marked as private and clearly went through a gated enclosure. This meant doubling back past the traffic police, who took no notice of me at all as I trudged past again, until I could find my way off the main road and out onto the plains. Although there was a maze of tracks marked on the map, none of them really went the way I was heading, so I simply walked due south using my compass.

The map suggested that there should be a gentle drop all the way from Teguital to Las Playitas. The reality was different. Between me and the end of the day's walk five kilometres away was wave after wave of pebble-strewn ridges that would have been almost unnoticed under normal conditions, but I was now very tired and my feet were dreadfully sore. I was desperate for confirmation of the end. At the top of each ridge I hoped to see the sea before me, but instead there was simply another barren little valley with a further ridge beyond.

The wind blew fiercely from behind me, threatening to blow away my hat, cooling but desiccating at the same time. All around was a Godforsaken barrenness, completely lifeless except for a reddish blush on the hillside here and there, where the succulent *Mesembryanthemum nodiflorum* grew in patches. This tiny plant seems to be able to find enough moisture to grow in such apparently waterless places that it would hardly be surprising to find it growing on Mars (Marsembryanthemum perhaps). Come to think of it, perhaps that is why Mars is red!

After almost an hour of trudging across desolate, sun-baked ridges I heard the cheery call of a Berthelot's Pipit in the distance, the first bird in all that time. I had come to the end of the barren lands, and was entering a cultivated area. A lizard darted across in front of me and there were Clouded Yellow butterflies and grasshoppers. Then I came to a road that I knew would take me down to Las Playitas, and all anxiety was gone. Just after reaching the road, there was a new house where an ugly looking mutt eyed me suspiciously. I fingered my dog-dazer nervously, but as I passed it simply turned round and slumped down in the sun, unable to muster the energy that would have been required to bother me. There is one advantage of travelling in the heat of the day.

Despite the fact that I was now walking downhill on a surfaced road, the last two kilometres to Las Playitas were the most painful of all. The blisters were causing me to walk oddly, as a result of which the muscles in my lower legs and ankles were now at the end of their tether. Consequently, I had to stop every two hundred metres or so to rest. At last, almost two hours after setting off from Teguital, I walked slowly and painfully through the doors of the Cala del Sol hotel into a smart, air-conditioned reception area where people were going about the business of being on holiday. I felt distinctly underdressed and out of place.

Nineteen

Las Playitas

I had arrived at the Cala del Sol hotel a day earlier than expected, and as a result it took some time for the helpful and efficient young German lady at reception to find me a room. Arriving a day early is probably something that doesn't normally happen here, given that everyone arrives by plane. So I waited in a plush chair in the foyer, desperate for a wash, a change of clothes and some food. It was a great relief when, after waiting for an hour, she finally signalled that a room had been readied for me, and I could at last have the luxury I craved.

The Cala del Sol is more than just a hotel, it is a tourist resort in its own right, and my room was almost at the furthest point, high up the hillside that separates the complex from the village of Las Playitas. Getting to my room involved a funicular railway as well as a combination of lifts and stairs. I had never heard of a hotel having its own funicular railway before, so this came as something of a surprise. Had I been in a better state I would have opted for the alternative of simply walking up the winding path, but I'm afraid that was out of the question for the time being. So I hobbled down the corridor to the 'station' where I waited with a group of smartly dressed Germans for a 'train' to arrive (it was clearly a German resort). I must have stood out like a sore thumb with my shuffling gait, heavy pack and dishevelled appearance, and I was embarrassed. I would hardly have been more alien to these care-free holiday makers if I had been dressed in a Gorilla suit. There was nothing I could do but act as nonchalantly as possible.

At last I reached the haven of my room, and what a haven! It was dominated by its windows and balcony with wonderful views over the bay and golf-course, and it had everything including, glory of glories, a large bath that I was looking forward to making full use of. But the most urgent action was to liberate my poor feet from socks and shoes. The sight that greeted me was a dreadful one. The twenty kilometres I had walked that day had raised huge blisters all over both feet. Some had burst and were weeping, others were just huge balloons of tightly stretched skin. As at Fustes, I had pushed myself to the limit and had reached haven at the last possible moment. Walking on the next day would have been foolhardy to say the least, and quite likely impossible. I had been justified in choosing a shorter route: I simply would not have made it if I had stuck to my original plans.

My spirits thus dented, I took a hot bath, or at least as hot a bath as my feet would allow. Then I plastered antiseptic cream onto my feet, and simply lay in the sun on the balcony for an hour or so. As hungry as I was, I simply couldn't face the thought of walking back across the complex in search of food. When I did eventually summon the energy, I discovered that all the clothes in my rucksack were unpleasantly wrinkled and certainly not the thing to wear in a hotel full of elegantly dressed Germans. I simply had to go to the laundrette.

Unluckily for me, the laundrette was at the opposite side of the complex. There was no option but to dress as well as possible and hobble slowly and painfully back to the funicular. I then had to walk past two swimming pools, three restaurants, the social club, diving centre and much more. But sadly, my efforts were poorly rewarded. The laundrette had washing machines but no washing powder, and an ironing board but no iron. The overpriced and under-stocked 'supermarket' next door did not sell washing powder, despite a sign on the door saying that it did, and the hotel's only iron had been lent to a man who, I was informed, would probably not bring it back until the following morning. So with a bag full of smelly laundry, I hobbled painfully and very slowly back to my room feeling pretty dejected.

It was getting dark by the time I emerged from the comfort of my room again, this time in search of a decent meal. Before eating I walked down to the beach. There was still a long way to go before the end of the walk, but I had at least made it from north coast to south coast, so I had achieved something. My early fears about the walk had been overcome, and the redrawn route had proved just possible. I didn't feel any sense of celebration, but I was confident of being able to continue the walk after a few days' rest. Over pizza and a glass or two of wine I considered possible routes for the remainder of the walk, and looked forward to a day of rest and recovery.

The following morning, my feet had recovered enough for me to walk down the hill to reception and, after a long, leisurely breakfast back up the hill again to my room. There was good news, too. I had initially been told I would have to move to another room for the remainder of my stay, but I now learnt that this would not be necessary. By lunch-time the washing problem had been resolved using some travel wash that I managed to buy, and the iron had reappeared so I was at last able to look reasonably presentable. I could now relax in my room with nothing to worry about and no pressures other than writing up my journal. I sat on the balcony enjoying the wonderful view and drinking a good bottle of Rioja. What a difference a day can make. I toasted Fuerteventura, applied more antiseptic cream to my feet and settled down to a gloriously indulgent afternoon.

★

Las Playitas could hardly be more different to Fustes and Corralejo. The resort is elegantly designed and landscaped, and exercise is clearly a high priority for many of those who stay there. The main swimming pool was divided into eight lanes and set aside entirely for serious swimmers, of which there was no shortage. They were collectively at it from dawn to dusk. No sooner had one group of swimmers finished powering up and down the lanes than another group took over. It seemed that people had come here for swimming holidays with professional coaches. And there was a phenomenal number of joggers too, including an organised run around the complex for juniors that finished at 8 a.m. This was evidently a place to come for a holiday based around physical activity

After Fustes and Corralejo, Las Playitas feels distinctly isolated. There is just one road in and there are no planes overhead. There are rocky hills separating it from the rest of the world that you would have to be either a nomad or a madman to attempt to walk across. I wondered which category I fitted into.

Twenty

Exploring the south

As at Fustes, having one day to rest was glorious, but the thought of another day doing so was unattractive, so for the second time on the journey I hired a car. The only vehicle available in Las Playitas was an automatic Chevrolet. It was grander and more expensive than I was looking for, but without a car I would be stuck in Las Playitas, so I took it.

Although Las Playitas itself is in one of the least interesting parts of the island for natural history, it is just fifteen kilometres from the wetland at Catalina Garcia (Plate 19). Although small, the lagoon is incredibly attractive to birds, and rare species turn up surprisingly often. Certainly, no birdwatcher should visit Fuerteventura without going there at least a couple of times. In April 1994, I spent a week on the island with a birdwatching friend checking details of sites for the *Birdwatchers' Guide to the Canary Islands*. Driving south from Tuineje, we noticed that a small lake had formed behind a new earthen dam.

Amongst the wetland birds we saw that day was a female Ruddy Shelduck, which proved to be the first record of this beautiful, rusty-coloured goose-like bird in the Canary Islands (Plate 5). Catalina Garcia Lagoon was clearly to its liking, because not only did it stay, but it was joined by others and they began to nest. Within a few years, it became one of the few common water birds on the island, and a very welcome addition to the island avifauna. It is numerous in Asia, but has only a very small, scattered population in north-west Africa, including the coast near Fuerteventura. John Mercer, who knew the opposite coast as well as he knew Fuerteventura, reported that up to sixty Ruddy Shelducks frequented the lagoon at El Aaíun. It seems very likely, therefore, that it colonised the island naturally.

★

Now that I had a car, a trip to the lagoon was my first priority. The track to the dam has deteriorated in the last few years because of the increasing number of birdwatchers who come here, and it was no easy matter to get the Chevrolet down to the lake. There is even a hide at the lake now, although it is generally locked and only available to a few local birdwatchers. In any case, the best way of watching birds here is still to drive across the dam, stopping at intervals to observe the birds from the vehicle.

On this visit there were thirty Ruddy Shelducks, more or less an average count here now. They are also numerous at Los Molinos reservoir near the west coast, and they turn up at many of the water tanks and any other areas of water that appear around the island. In fact, there may be more Ruddy Shelducks in Fuerteventura now than there are in the whole of Morocco, where it remains very scarce. They are certainly very attractive birds, but peace-loving they are not! They seem to spend an inordinate amount of time posturing and chasing each other around the lake, cackling and wailing noisily as they do so.

Another species that has benefited greatly from the creation of the lagoon is the Black-winged Stilt. This most elegant black and white wader with its crazily long red legs was a rare visitor to the island in the past, but is now regular both here and at Los Molinos. In the last few years, small numbers of Spoonbills have become regular in winter, and colour rings on their legs confirm that they originate from a breeding colony in France.

The ability of the lake to attract the rare and unexpected is extraordinary. On this occasion there was a male Ring-necked Duck, a rare visitor from North America. There were also migrant birds pausing here on their way north towards Spain and other parts of Europe from wintering areas south of the Sahara, including a Red-rumped Swallow, a very handsome Woodchat Shrike and two Spotted Crakes. Spotted Crakes normally skulk in reeds or other dense cover, and are hard to see. Here though, there was nothing for them to hide in, so I had great views.

As well as its importance for birds, the lake is also one of the key sites in Fuerteventura for dragonflies and damselflies, of which just seven species have been recorded. The bright red Scarlet Darter seems to be the commonest species, whilst larger dragonflies seen here are likely to be either the Emperor or slightly smaller Lesser Emperor. The little damselfly found on the island is the Saharan Blue-tailed Damselfly, which is very like the familiar Blue-tailed Damselfly of Europe.

★

From my balcony in Las Playitas I could see the mountain of Montaña Cardón (694m), an isolated peak in the sparsely populated south-west corner of the main part of the island. There was something about its isolation that drew me to it, and I had hoped to include it in my walk. Furthermore, it is home to a number of rare plants. Reluctantly, however, because of the state of my feet I had now decided to reduce the distance I would walk as much as possible. A diversion from the south coast to include the mountain would involve an extra day's walk. On top of that, due to its isolation, I would have been weighed down with as much as I could cram into my already bulging pack because there would be little chance of buying supplies during more than a day's walk. Although I now accepted that I could not include Montaña Cardón in the walk, I was still very keen to climb it as an excursion from Las Playitas.

I wanted to rest my feet for another day before attempting the climb, but was also keen to visit the slopes below Montaña Cardón's eastern cliffs, where I was fairly confident of finding a succulent plant that I had not previously seen on the island. The desk calendar I had been given in Fustes included a picture of a wonderful stand of *Euphorbia canariensis*, a cactus-like succulent spurge that grows in massive clumps of three metre high organ-pipe stems. As its name suggests, it is endemic to the Canary Islands. It is common on some islands, but very local in Fuerteventura. Up until then, I hadn't known where to look for it, but the local name for the plant is *Cardón*, so it seemed likely that it grew on the mountain of the same name. Now, despite the fact that the mountain was many kilometres away, with the aid of my binoculars I could see blue-green clumps on the slopes below its cliffs that I was convinced were this species.

In the late afternoon I drove to Cardón, a small settlement that also takes its name from the Euphorbia. It was a rough scramble from there up fairly steep ground and across several deep rock-strewn gullies, but after twenty minutes I reached my goal. The clumps were up to five metres across, and so dense that no animals could graze within them (Plate 20). For this reason they sheltered a number of plants that were entirely absent from their barren surroundings, including the Madder *Rubia fruticosa*, with its whorls of prickly leaves, and the scrambling asparagus *Asparagus pastorianus*, with its thorns and red berries

I sat down between the Euphorbia clumps and enjoyed a moment of great satisfaction. In all my visits to the island I had never previously encountered these plants, and, now that I was sitting amongst them, it was almost as if I had just confirmed that dragons still existed. Despite centuries of destruction and grazing, these magnificent plants live on, in much reduced numbers admittedly, but still in fine stands covering a fair stretch of the hillside. Above them, the almost sheer cliffs of Montaña Cardón formed a splendid backdrop.

Euphorbia canariensis was important for the pre-Conquest islanders, who used its dried stems as tinder for making fire. Over centuries, this must have greatly reduced the size of its colonies. In more recent times, and in common with the other species of *Euphorbia* on the island, it has been collected for fuel. Due to all these pressures, it is now only found on the slopes below Montaña Cardón and in a few remote places in the Jandía mountains.

★

Before returning to the hotel, I drove on past Las Playitas to the lighthouse at Entallada. The road itself is narrow with steep, unprotected drops on the final ascent to the lighthouse, and definitely not for those of a nervous disposition. The lighthouse is an attractive building, certainly the finest one on the island. Unlike the others, which are all at sea level, this one is on top of a 150m high sea cliff. The central tower, which houses the light, looks not unlike the towers of the grander churches on the island. Then there are lower towers at the ends of the building which mirror the front of the Casa de los Coronelles in La Oliva. The corners, base and surrounds to the shuttered windows are of neatly squared dark stone, and the walls have the appearance of delicate white lace, due to the heavy mortar fill between the rough stones. Added to this, the setting makes this spot worth visiting in its own right. The views of the coast from the cliff top are splendid, and inland there are stark mountain slopes, dominated by the sharp peak of Vigán, only 464m high but looking much higher and immensely forbidding.

★

Having studied Montaña Cardón from my balcony, I decided that the easiest way to climb it was from the south. The road itself climbed to about 280m, from where there was a steep ascent to the southern end of the ridge, then a one and a half kilometre ridge walk to the summit.

After an early breakfast, I drove to the mountain and parked by the highest point of the road. From there it was a steep ascent over very loose rock, passing three magnificent palm trees. Looking back, there were fine views over the palms across the south of the island, with the sea in the distance (Plate 31). But I now turned my attention to the mountain. There seemed to be no easy way up the mass of jumbled rock before me, and it was so steep in places that it was virtually a cliff. Without doubt, I could not have managed the climb with my heavy pack, but having studied it for a while, I thought I could see a way up. At least I was determined to have a go.

As I began the perilous climb, a magnificent view of the whole of the north coast of the Jandía peninsular began to open up across the gaping chasm to my left (Plate 32). In the foreground were the sandy wastes of the La Pared isthmus and beyond, the high mountains of Jandía, with Pico de la Zarza towering above all others, at 812m the highest point on the island. But I needed my full concentration to find a way up the rocks to the ridge, scrambling on all fours most of the way and climbing up sheer rock in places. It was more than a little scary, and the persistent mewing of a Buzzard circling above me seemed quite menacing in the circumstances.

At last, a little under an hour after setting out, I found myself on the ridge where the wind was pleasantly cool. Looking back the way I had come there now appeared to be nothing but a sheer drop, so it seemed best to concentrate on the ridge walk that now stood between me and the peak on Montaña Cardón, and not think about the descent!

Although the ridge walk had looked easy from my balcony in Las Playitas, as is so often the case, now that I was on it, it didn't look that simple. In places the ascent was gentle enough, but there were several rock bastions to contend with. Admittedly, none looked too difficult, but there was a substantial wall built along the whole length of the ridge that looked as if it might make climbing some of the bastions more complicated. The wall was two metres high in places and built with large rocks. Presumably it was put there to stop goats moving between one estate and another, but the effort involved in building it up here must have been enormous. It is difficult to imagine the scale of the labour that would have been involved (Plate 33).

It took half an hour to gain the true summit, a flattish, insignificant point on the ridge. There were wonderful views. To the west I could see the whole of the Jandía peninsula. Away to the north I could make out the viewpoint at Morro de Velosa where I had slept under the stars on my way south. Further still and almost thirty kilometres to the north east, I could even see the peak of Rosa del Taro which I had climbed on my way to Fustes. And below were the dull green patches of the *Euphorbia canariensis* that I had been so thrilled to visit the day before.

But I must admit that I was more than a little disappointed. I had seen none of the rare plants that this mountain is known to support. Also, I now realised that the caves and ancient fortifications I had read about in John Mercer's book were on the jagged peak of Espigón de Ojo Cabra that now dominated the view to the north. This fearsome spike of un-climbable looking rock was separated from the Cardón ridge on which I now stood by a deep cleft. It is the Espigón de Ojo Cabra that has the caves and fortified summit, and is reputed to be one of the last places where the pre-Conquest islanders took refuge before the island was entirely subdued by European settlers. According to Mercer, there are twenty or thirty caves hollowed out of the soft rock near the summit, in which the islanders had survived for a while after Europeans arrived. At the time of his visit in the 1960s, the ground below them was full of decorated potsherds and limpet shells, although I suspect these will now have gone. The summit had been fortified with walls to prevent the European settlers from climbing up to them.

After I had taken a few moments to enjoy the view, I headed back along the ridge. Looking back, I was amazed to see two people standing on top of the Ojo Cabra rock pinnacle. Clearly it is not as un-climbable as it looks, and even if the summit itself is too difficult for most, the caves and fortifications must surely be worth a scramble to visit.

The walk back along the ridge was uneventful except for a light shower of rain, one of the very few in the three weeks of the walk. However, the descent from the ridge was a stiff challenge and there were sections of vertical rock where I had to take great care: going down always seems to be more difficult that going up. There was a real sense of relief when the vertical sections ended, the slope finally eased, and I could finally walk upright again. The three stately palm trees came back into view and comforting sounds of chirping sparrows and cooing doves drifted up towards me. Amongst the rocks, I could hear the bizarre buzzing of Trumpeter Finches and the calls of a family of Fuerteventura Chats. Overhead the same Buzzard's mewing now sounded triumphant, whereas earlier it had sounded menacing. Sounds influence our state of mind, hence the dark music that plays when something dreadful is going to happen in a film, but how often are we aware that our own state of mind influences the way we hear sound?

The Buzzard in question evidently had a nest nearby, perhaps in the tall palms, and it seemed none too pleased with my intrusion. It started to mob me in a way I had never witnessed before, and I could hear the wind rushing through its flight feathers as it dived to within ten metres of me. Although I had never heard of Buzzards attacking people, I must admit that I was slightly perturbed, and I picked up a dead Nicotiana stem that was lying handily about the hillside, just in case it got any bolder.

Three hours after I had set out I reached the car and put the stick in the boot. By chance it was just the right length to serve as a walking stick, and I thought it might be useful the following day, when I would be resuming my walk down the island.

TWENTY-ONE

The problem of Jandía

After climbing Montaña Cardón, I had half a day left, with nothing in particular in mind. I found myself driving west towards the Jandía peninsula. Jandía was originally a separate island. What we see today are the eroded remains of what used to be a significantly larger volcanic island to the south-west of Fuerteventura, now joined to it through a combination of more recent volcanic activity and the deposition of vast quantities of sand. Because it was once a separate island, Jandía has its own endemic plants. The most exotic of these is *Euphorbia handiensis*, which, like *Euphorbia canariensis* is cactus-like. Though not as large, it is nevertheless a handsome plant, forming dense, spiny, clumps. Sadly, many have been taken for horticultural purposes in the past and it is now only found in a few scattered colonies on the slopes in the south-west of the peninsula.

★

According to early historians, in pre-Conquest times there was a separate tribe living on Jandía. They were separated from the tribe on the main part of the island by a wall, which was built across the isthmus near the small resort of La Pared. Indeed, La Pared means 'the wall'. It was about a metre wide and slightly more in height, but most has now gone, the stones having been taken away for building.

Until the early 1980s, Jandía was isolated for another reason. It had no sealed road to connect it to the main part of the island. The surfaced road stopped at Tarajalejo, and it was a gruelling thirty-five-kilometre drive down a badly rutted dirt track to the little cluster of hotels near the lighthouse at Morro Jable. In a hire car it was a matter of picking a route carefully over the rough surface, fearful of the effect of pot-holes and rocks on tyres and sump, but the white Mercedes taxis took no notice at all of the conditions. They simply roared along, taking wealthy German holiday makers from the airport to the splendidly isolated luxury resort at Stella Canaris, creating great clouds of dust. As if by magic, just before the hotels there was a kilometre or so of beautifully surfaced road.

The reason for the early development of tourism in Jandía is obvious enough. It has a super-abundance of sand. The whole of the south coast between Costa Calma and Morro Jable, a distance of some eighteen kilometres, is one glorious long beach. Since the road was finished more than twenty years ago, the construction of hotels and tourist facilities has been non-stop. The result, of course, is that the character of the peninsula has completely changed. What was once a harshly beautiful place, with more than a touch of adventure about it, is now an ordinary tourist resort. Where once there were wide open spaces, there are now apartment blocks and landscaping so dense in places as to almost form woodland. However, as well as making the place look more attractive, the landscaped areas are also a real boon for migrant birds.

★

Although I hadn't planned to do so, I found myself driving to the very end of the peninsula, more because there seemed to be nothing to detain me along the way than out of any desire to reach the end. Even now that there is a good road as far as Morro Jable, the twenty kilometres from there to the end of Jandía is still a dirt track, and a journey not to be undertaken lightly in an ordinary car, not least because your hire car insurance won't cover you if anything goes wrong. It is not an inspiring drive either, just an endless series of dusty bends, the track rising and falling gently over the flattish land between the central mountain spine and the south coast. There are no beautiful beaches and the mountain slopes are gentle on this side: the spectacular cliffs are all on the eroded northern side that is more difficult to access. To drive at more than thirty km/hour on the rutted surface would be madness, though there are plenty that do, and yet there is an endless procession of vehicles going to the point, simply because it is there. Finally, reaching the end, there is nothing at Punta de Jandía that you can't see in a thousand places elsewhere in Fuerteventura. Except, that is, for the lighthouse, but even that has no architectural merit, certainly nothing to compare with the splendid building on the cliffs at Entallada.

In the early afternoon I found myself sitting on the black rocks at the point, watching the waves crash into a pretty cove just north of the lighthouse. An Oystercatcher, a rare bird in these parts, flushed from the small beach and flew away up the coast, a flash of black and white against a vivid blue sea. A hundred years ago it might have been a Black Oystercatcher, a species unique to Fuerteventura, Lanzarote and adjacent islands, but sadly this bird is now extinct. It was entirely black with red bill, eye and legs. The last confirmed record was on 3rd June 1913, when David Bannerman described his 'intense joy' at having shot one on the little island of Graciosa. Of course, he had no way of knowing then that no birdwatcher would ever see the species again, and I suspect and hope that he would have been just a little embarrassed if he had. His was an age when the world seemed a great deal larger than it does now, and it probably seemed quite impossible that shooting a few specimens could cause extinction. Fortunately, there are virtually indistinguishable black species of oystercatcher inhabiting the rocky coasts of sub-Saharan Africa (which the Canary Island birds were undoubtedly descended from) and Australia, where I had been lucky enough to see some in 2006.

★

Now that I was at the far south-west point of the island, I had to confront the question that had been in my mind ever since the planning stage. Where should the walk end? From a geographical perspective, it seemed obvious that it should be here at Punta de

Jandía, the point of the island furthest from Corralejo. The problem was that I didn't really want to walk the twenty-kilometre dirt track from Morro Jable to the point. The thought of plodding down that awful track with a constant stream of vehicles passing me and covering me in dust was not a happy one. Furthermore, having reached the end, I would then be faced with walking all the way back again unless I could get a lift. I was doubtful that tourists would want to give a lift to an oddball with a rucksack.

It would undoubtedly be a wonderful feeling to have walked the whole length of the island. But did I have enough willpower to tackle this last, most difficult challenge? I had a few days left before I would finally have to make the decision, but at that moment I had a feeling that Morro Jable would be the end of the walk.

★

The following morning I woke at six, with a brilliant full moon flooding my hotel room in Las Playitas with light. Today would be the start of the next phase of my journey: a three-day walk along the whole of the south coast and deep into Jandía. Indeed, if I decided to stop at Morro Jable, this would be the final stage. But first I had one last journey to make in the Chevrolet. The previous day I had booked myself into a smart-looking hotel in the resort of Tarajalejo, which was to be the next stopping point. I now put the few things I would need for the day's walk into my day bag, and put everything else in the rucksack. Then I drove to Tarajalejo and left the rucksack with the friendly staff at the reception desk. Today I would be travelling light. The sixteen-kilometre walk along the coast would be over a series of steep ridges and I knew it would be much easier without the rucksack. Having completed the round trip to Tarajalejo, I tucked into a hearty breakfast.

Looking out from the breakfast room, the path west from Las Playitas was etched like an inverted question mark on the steep hillside across the bay, a faint dusty mark on black basalt. Each morning at breakfast I had watched some of the more fitness-conscious residents scrambling up this path. The time for me to take that path had come.

Unfortunately, I wasn't feeling very confident about the day's walk. This was partly a fear of the unknown, since the route was through a part of the island I was not familiar with. I knew it would be a hot, strenuous walk over rough ground, with no paths and many steep hills. Perhaps, also, it had been unwise to drive to and from Tarajalejo that morning, because this had reminded me how desolate and hilly the terrain was. Because of all the valleys I would have to cross, I estimated that the total amount of climbing would be in excess of 1000m. As ever, the biggest risk was dehydration, and I was determined to carry enough water to get me through. Unfortunately, because my day-sack was so small, I didn't have enough room in it for the water I needed, so I would have to carry an extra bottle in my hand. Still, if I found myself defeated by the walk, I could throw in the towel at either Gran Tarajal or Giniginamar and simply get a taxi to Tarajalejo. I would reassess the situation as the day unfolded, and decide whether I had the strength and commitment to carry on. At least I would be travelling light, there was the possibility of obtaining additional refreshments along the way, and my trusty stick from yesterday's encounter with the Buzzard would help negotiate the difficult terrain. I paid my bill, retrieved the stick from the flower bed where I had hidden it the day before, and headed west.

TWENTY-TWO

Gran Tarajal

By the time I set out for Tarajalejo on the next leg of the walk, the early morning hill-walkers from Las Playitas had all returned for breakfast. Hence, as I started up the steep slope beyond the hotel gardens there was nobody to be seen, but for some reasons there were goats racing about on the ridge above. From the top of the hill the town of Gran Tarajal was just visible through a gap in the hills, but there was no obvious way of getting to it. In front of me there was a very steep drop that I didn't like the look of. I wasn't sure whether to head inland, following the higher ground, or to hug the coast and hope to find a path. I initially chose the inland route, picking my way carefully over the rocky ground, but it seemed to be heading in the wrong direction.

After a while I noticed a path of sorts heading towards Gran Tarajal from the cove at the mouth of the barranco below me. I turned round and headed back towards the sea cliffs, finding a track of sorts where the goats were evidently in the habit of descending almost vertically to the cove below. However, it took no account of human abilities, and I ended up finding my own way down the scree slopes, my stick already coming in handy by providing stability where I would otherwise have had difficulty staying upright.

A young couple was camping at the isolated cove, the aptly named Playa de los Pobres, or Poor People's Beach. No simple tent this, but a minor settlement with tarpaulins stretched across the deeply incised end of the barranco to provide shade. Unfortunately, the gully down which I approached the camp evidently served as their latrine, and as I approached, a man was squatting with his back towards me, unaware of my approach. In order to spare his blushes, I left the gully and struck out across the hillside, but nevertheless he had quite a shock when he saw me, and scuttled awkwardly for cover. Unfortunate for him but most amusing for me!

Beyond the cove I soon came to the path, and within an hour of starting out, I came to the top of a pass and looked down on Gran Tarajal. As I started down the hill towards the town, a group of four bulldogs appeared on the track in front of me. It was an unusual sight and not a welcome one. Fortunately, they seemed keen to avoid me, and trotted off rather comically on stumpy legs towards some houses away to the right. I gave the houses a wide berth, and soon found myself in the big tamarisk-lined barranco, the Río Gran Tarajal, that leads down to the town. Sadly, it has always been something of an unofficial rubbish dump.

For some reason I was reminded of my entry into the village of Lajares on the first day, and how uncertain I had been then. In contrast, although I knew I had a tough day's walk ahead of me, I was confident now. Reflecting on what I had already achieved, I was quite proud to have made it so far, and very grateful not to have given up on the second day.

★

Back in February 1981, the day after our car failed to start on the plains by Tiscamanita, we had had the same problem again. On that occasion we had parked for the night in a quiet place along the Río Gran Tarajal, about four kilometres north of the town. Having failed again to start the car, we were just contemplating our next move when, to our surprise, a venerable old black car came down the barranco towards us. It stopped and a wonderful old man and his wife beckoned us in with broad smiles. It seemed they were on their way to the market in Gran Tarajal and were delighted to be able to give us a lift. And a perfectly normal lift it was in all respects except one: our old friend had made no use at all of the roads, but brought us to the dusty car park in Gran Tarajal through a series of drains and culverts, some hardly big enough for the car to pass through. Presumably he had no tax and insurance, so preferred not to use the road, a wise enough precaution given that he would have had to pass the police station on his way into town.

★

As I made my way down the barranco towards the town, Sardinian Warblers scolded from the tamarisks and there was the now familiar cooing of Collared Doves. In the past, this barranco was filled with the murmuring of Turtle Doves, but it looks as though they are rapidly being replaced by their bolder, more adaptable relatives.

At the bottom end of the barranco, where it skirts the town, a not inconsiderable amount of money has recently been spent to turn it into urban green space. The tamarisks that gave Gran Tarajal its name (*tarajal* is Spanish for tamarisk) have been removed. On the eastern side of the barranco, where a new residential area is being built, there is a 'fitness park' that goes on for some hundreds of metres, with all manner of wooden exercise equipment. To what extent it is actually used I couldn't say, although there was certainly nobody there when I passed it. I couldn't help wondering whether this was really just a way of tidying the area up, or perhaps a way of spending some money on the place. And how long will it last before it looks dishevelled and abandoned?

★

At the entrance to the town, sandwiched between the petrol station on one side and the denuded barranco on the other, is the finest public open space in the town. After the desolate hills, the lush green lawns and palm trees with their cool shade were much too seductive to pass by. I chose the bench that had the deepest shade. Above me the palms rustled lazily in the breeze and Spanish Sparrows chirped and chattered enthusiastically.

Gran Tarajal cannot claim to be pretty: it is a working town, not a tourist resort, and as such it gets little attention in the guide books. Even the AA Spiral Guide, which is one of the better ones simply says:

> "Although it is the second biggest town on the island after Puerto del Rosario, there is little to attract visitors to Gran Tarajal aside from a large well-kept black sand beach."

The beach is gritty rather than sandy (though it is indeed kept clean) and the promenade is unpretentious. The bay on which Gran Tarajal sits is dominated at its eastern end by a rather severe black lava cliff which is topped by a cluster of masts. In the past, local people were in the habitat of disposing of worn-out cars by pushing them off this cliff into a watery graveyard below. At the eastern end the seafront is marked by an austere harbour wall.

The outskirts are an ugly sprawl of industrial estates. No doubt many tourists see these and avoid coming into town at all. They are not missing much, but it is a significant town built around the commerce of its small port. However, Gran Tarajal does have one significant attraction: it is the only centre for miles around with a good variety of shops that don't charge tourist prices.

Visitors to Gran Tarajal do need to be aware of one thing. Though it is not a large place, it is a maze of narrow one-way streets, especially the main residential area stretching up the hill. Indeed the one-way system is so confusing that some years ago I received a parking ticket when I parked facing the wrong way round in a one-way street. This was more than a little inconvenient at the time, because paying the fine involved finding a particular building in Puerto del Rosario. So if you are driving into Gran Tarajal, my advice would be to park near the public park. It is only a small place anyway, so nothing is very far away.

★

Gran Tarajal was uninhabited until the nineteenth century. It was too prone to attack by pirates, who were a constant problem in the seventeenth and eighteenth centuries. In 1740 English pirates turned up not once but twice in the same year! At that time there would have been nothing to loot at Gran Tarajal, but it provided a convenient anchorage for a raid into the island. On both occasions the pirates took the obvious route inland – up the tamarisk-lined barranco, which was presumably pristine and not full of rubbish in those days (and definitely no fitness park), and found themselves in the town of Tuineje, fourteen kilometres to the north. And on both occasions they suffered the same fate: slaughter or imprisonment. Reasonable enough, I suggest, in the circumstances.

In the late nineteenth century a small settlement was established here as a stopping point for the inter-island steam ships and that is why the town exists. The harbour was too shallow for steamers to draw up alongside the quay, so they anchored offshore and transferred goods and any passengers by means of lighters. David Bannerman gives a brief account of proceedings during his journey to the island in May 1913:

> "We were awakened next morning at 5.30 by the sudden cessation of the engines, and thinking we had arrived at our destination [Puerto del Rosario], we scrambled up on deck to find we were lying off a tiny port called Gran Tarajal. It was a glorious dawn, and the island looked most alluring in the early morning light. There appeared to be only about two houses in sight, which were built on the beach at the entrance to an attractive valley or barranco, which led into the interior of the island. The bottom of the barranco was thickly lined with tamarisks, but the sides, which looked steep and rocky, seemed absolutely bare of vegetation, as indeed was most of the land in sight. In the distance, beyond the tamarisks, a few clumps of palms added a picturesque note, which they always do to any scene, and above their feathery heads a low range of blue hills stretched into the distance.
>
> A heavy surf boat was putting out from the shore as we

came on deck, and we made out a string of camels kneeling patiently on the beach – apparently waiting for the goods which, with much creaking and groaning of the winches, were being hauled out of the hold of our little steamer."

The town obviously grew rapidly over the next fifty years or so, because in the 1960s John Mercer described the view of Gran Tarajal from the steamer as "a jagged grey mass of half-finished houses. Rectangles of undisguised cement pocked with small square windows, they spread away up the bleak mountainside." Henry Myhill was hardly less dismissive of the town in his 1968 book on the Canary Islands. The best he could say about it was that "the only thing which is great about Gran Tarajal is its unattractiveness". This would be a good description of the town as I knew it in the early 1980s. However, by then the inter-island ships no longer stopped here. The larger vessels that had been introduced by then needed a proper port and one was duly constructed further down the coast at Morro Jable, previously an isolated fishing village. However, this has by no means stopped Gran Tarajal from gaining in prosperity and importance. Today it is expanding more rapidly than ever, and is a thriving commercial centre.

Although the ferry no longer stopped at Gran Tarajal during my time on the island in the early 1980s, it was visited, from time to time, by the Spanish navy. I will never forget the first time I saw the ageing warships anchored some way offshore in the deep water of the bay. There was an aircraft carrier and several other big naval ships, but they were in such a poor state, with peeling paint and rusting hulls, that the effect was frankly somewhat comical. How welcome the sailors would have been in Gran Tarajal I am not sure, but hopefully rather more so than the English and Berber pirates whose ships moored here from time to time two hundred years earlier.

★

I found a café on the promenade for a cup of coffee, and then carried on towards the harbour. The local fishing fleet is based here, and there were lots of traditionally painted wooden fishing boats. In the dry dock there was a fine example, with the rather surprising name *Horatio*, though presumably not after the British Admiral Horatio Nelson, who in these parts is remembered as the villain who attacked Santa Cruz de Tenerife in July 1797. A modern lifeboat is also based here.

At the base of the quay is perhaps the most significant building in town: the Cofradia de Pescadores, the Fishermen's Association of Gran Tarajal. And it is not only for the fishermen of Gran Tarajal, but also those from La Lajita, Las Playitas, Ajui, Tarajalejo, Giniginámar and Pozo Negro, all the fishing communities in the south. It was noon when I went by, and clearly all the fishing was over for the day. As I passed, the sound of animated discussion drifted out from the bar.

Until recently, Gran Tarajal ended abruptly at its western end, where the rocky hill comes right down to the harbour. The roughest of tracks led round the promontory into the valley beyond (Tuna Valley). Now the town has spilt out beyond this natural barrier and there is a proper road passing the harbour to the bottom end of the valley, where the 'Vista Mar' residential area is now being constructed.

Beyond the new development area there was no trace of any path leading west. Before me was a steep, very rocky slope. I had a tough walk ahead of me.

Twenty-three

A cliff walk

Although most of Fuerteventura is technically semi-arid rather than true desert, this is not the case along the south coast. Here, rainfall is comparable to parts of the Sahara averaging below 50mm a year, which is less than a quarter of the amount that falls in the mountains. While the heat is nowhere near as intense as it is in the Sahara, and the humidity is higher, the lack of rainfall explains why the hills of the southern coastal fringe are so bare. They seem to be utterly without vegetation of any kind, even when scarce winter rains have been sufficient to encourage a flush of new growth in the rest of the island.

In fact, close examination of the hillside above Gran Tarajal that I was now standing on revealed that there was a thin scatter of minute annual plants: a tiny, white-flowered asphodel here and there, or a lone *Echium* holding its purple flower just a few millimetres above the bare soil. There were few birds either, not even the ubiquitous Berthelot's Pipit, just the odd Raven or Kestrel patrolling the hills.

Dust-dry rocks present a daunting prospect for a day's walk at the best of times, and when they are piled loosely on a steep slope with no hint of a path to follow, the effect is doubled. The hill guarding the western side of Gran Tarajal is not really that high, less than 300m in fact, but the ground conditions made it a formidable prospect. I also knew that once I had found a way over it, I would be faced with a series of similar hills all the way to the hotel in Tarajalejo about ten kilometres to the west.

The steep hillside I found myself on was a mixture of large rocks and loose scree, though thankfully it all seemed reasonably stable, and in the event I was able to pick my way up the flank of the hill without too much trouble. Behind me there was a wonderful view back to Gran Tarajal and its harbour, a mass of whitish buildings huddled around the bay between desert hills and blue sea. From the top I looked west over a series of barren ridges, with the mountains of the Jandía peninsula on the horizon beyond them. The good news was that all the ridges before me seemed somewhat lower than the one I was standing on.

To my surprise there were about twenty vans and tents arranged neatly around the remote cove below. At the time I thought perhaps it was a hippy community, but during the next two days of the walk I came upon the same phenomenon at every cove that had any kind of vehicle access. The people of Fuerteventura were enjoying their Easter holiday on the undeveloped southern beaches where tourists rarely venture. The clusters of vans were family groups, each complete with grandparents, numerous children, pet dogs and inflatable boats. And a happy time they seemed to be having.

I found a rock on the hillside overlooking the cove to eat a meagre lunch. For the first time on the walk, small flies were a nuisance. They appeared to be sweat flies, perhaps attracted to the area by the campers. Although heat and thirst are problems to contend with on any walk in Fuerteventura, walks on the island are, on the whole, blissfully untroubled by insects and creepy-crawlies. After eating, I continued down the steep slope to cross the wide-bottomed barranco about 500m above the camp. The thump of music drifted up towards me from the camp, and a motorcycle roared down the track towards it. Then I started up the next slope and was soon back in the peace and tranquillity of the hills.

On a rocky slope I noticed an unfamiliar plant with somewhat fennel-like leaves and small white flowers. From the description I made of it, I later identified it as *Crambe sventenii*, a rare member of the crucifer family found only in the dry rocky hills of southern Fuerteventura.

The next barranco was utterly unspoilt. The dry stream bed showed no trace of vehicle tracks, and there were no ruins or other signs of human interference. Even here, though, there were a few people camping rough in the bay. They were fishing from two rubber boats, which probably explains how they got there.

At last there were a few birds too: a pair of Fuerteventura Chats hopping from rock to rock and a Barbary Partridge calling from the far hillside. For some reason I was reminded of the day in 1979, during the Fuerteventura Houbara Expedition, when Paul Goriup and I walked the barrancos east of Las Playitas looking for the chats. It was not always easy to get from one barranco to the next, and at one point we had decided it would be easier to wade round a headland rather than clambering over the ridge. We took socks and boots off and put our binoculars away in our back-packs. Paul went first and made it to the next beach without too much difficulty. I followed gingerly after him: I was a very poor swimmer at that time and wasn't too sure I could get round. At the deepest point the water was up to my chest and I was having some difficulty holding my boots above the waves. Paul encouraged me to throw them to him so I could move more easily. Unfortunately, my aim was poor and they somehow looped over my head and landed in the waves further out! I lunged awkwardly forward, grabbed the boots and somehow made it back to dry land in what seemed like one interminable moment. Not surprisingly, Paul found the whole incident hilarious: of course he had no idea that I couldn't really swim. It was only later, when I was 'teaching' my daughters to swim, that I gained some confidence in water.

★

I hauled myself up onto yet another ridge and dropped down into a remote barranco that was as unspoilt as the last. Here there was nobody camping by the tiny beach, so I stopped to enjoy the tranquillity. It had now clouded over and the temperature had dropped significantly. It was perfect walking weather in fact.

Looking at the bed of the barranco, I noticed that, at some time in the past, there had been a flood of sufficient strength to cover the barranco bed in a thick deposit of grit and small stones. This layer was higher in the middle where the full force of the storm water had carried the fragments of hill seaward. It was all the more remarkable since the barranco is barely more than a kilometre long. Because of the lack of human disturbance, it looked as though the stream had just dried up, yet in all probability there had not been a flash flood in the barranco for years.

In several of the barrancos I noticed the large, fleshy leaves of a plant I have come to know as the 'Ginny' Plant, for the simple reason that it is quite common in the vicinity of Giniginámar (Box 19).

★

It was early afternoon when the village of Giniginámar finally came into view below me, its white buildings scattered around the mouth of a wide valley. The low houses of the original fishing community huddle close to the pebbly beach at the eastern end of the cove, tucked sensibly under low cliffs out of the predominant wind. Scattered around the far side of the bay was the inevitable cluster of campervans. Just inland on the western side of the valley, standing separately from the village, is a small, low-key tourist complex that seems to be struggling to survive. The lack of a proper sandy beach has allowed it to preserve an air of tranquillity: a rare thing indeed on this island.

It was less than straightforward to find a way down into Giniginámar. The descent was fearfully steep and covered in awkward loose material. I was very glad of my improvised walking stick that had served me well as a third leg all day. I made my way to the little Bar Olas del Sur (the Southern Waves), on the beach below the cliffs, where I found a table on the cramped veranda. From here I could gaze out to sea while eating a good meal. But now that I had stopped, I realised that I was very tired. My hands and feet were tingling slightly from the exertion and my energy levels were really low. The steep, rough ground had certainly taken its toll.

★

24 – Left: The Estrella del Norte roundabout sculpture on the way out of Corralejo.

25 – Below: Goats were the mainstay of the island economy until the tourist boom.

26 – Bottom: La Oliva church with its muscular tower.

27 – A six-metre deep gully, one of several between Lajares and La Oliva where water has worn through the crumbling calcareous crust that covers this part of the island.

28 – The restaurant in La Matilla with Muda in the background.

29 – Church and square, Antigua.

30 – The old cathedral at Betancuria.

31 – *Above left: Palm trees on the slopes of Montaña Cardón.*

32 – *Below left: Climbing Cardón there were spectacular views of the Jandía Peninsula.*

33 – *Above: The ridge leading to the summit of Cardón, with the Espigón de Ojo Cabra to the left.*

34 – *Right: Sculpture of Aulaga on the roundabout at La Lajita.*

35 – Red-vented Bulbul.

36 – The last campsite overlooking the coastal lagoon on the south coast of Jandía.

37 *Bertelots Pipit is common throughout the island.*

38 – Southern Grey Shrike.

39 Raven.

40 Ground Squirrel.

Over a leisurely lunch I used my binoculars to scan for seabirds. There were a few Cory's Shearwaters miles offshore and only just visible, and a single Gannet. Gannets are big seabirds (the wing-span is six feet) that nest on sea cliffs and rock stacks around Britain and Ireland in particular, where some colonies are tens of thousands strong. They winter at sea, mainly off the West African coast, and can often be seen from the coast of Fuerteventura.

Various rare Shearwaters and Petrels nest amongst the islets to the north of Lanzarote. However, they are rarely seen from Fuerteventura because they come to land only after dark, and spend the day feeding far from land. There is perhaps a better chance of seeing one of the tropical species that occasionally ventures further north than normal, such as the exotic Red-billed Tropicbird, the adults of which have half metre long tail-streamers. About a hundred pairs nest on the Cape Verde islands, 1,000 miles to the south.

★

Just down the road from the bar, on one of the old fishermen's cottages, was a mural with a nautical theme. Dark grey top shells and other, lighter coloured shells had been used to create a picture on an immaculately whitewashed wall. It was simple art, nothing intricate, a fishing boat and an upright structure with a circle and a diamond perhaps representing a marker buoy, framed on either side by additional blocks of shells.

I was now looking forward to the six-kilometre cliff walk between Giniginámar and Tarajalejo, which is in the Sunflower walking guide, although I was slightly daunted by its warning. It describes the path as 'rough and vertiginous – only recommended for adventurous, surefooted walkers with a head for heights'. I reckoned I could probably fit into that category, but exactly how vertiginous did it mean?

To get onto the coastal path, I had to find a way through the little knot of whitewashed buildings at the western end of the bay, which are still occupied. I wasn't sure whether the path skirted round the back of them or passed between the narrow gap between them. I wasn't sure either, that I liked the look of the fifth bulldog of the day, which was standing proprietarily by the path. Fortunately, I spotted an elderly man who evidently lived there and asked him which way the path to Tarajalejo went. He was pleased to help, and showed me through the gap between the houses, talking busily away all the time, much of which I didn't catch. The bulldog growled and snorted at me, and strained against its tether, but I needn't have worried. The rope was too short for him to reach the ankles of anyone walking along the track! Anyway, by engaging my resident guide, I had nullified his advantage, and the dog-dazer remained unused as ever.

The man, who had presumably lived there all his life, was astonished that I should want to walk to Tarajalejo. "Why not get a lift along the road" he asked. No I wanted to walk to Tarajalejo, I explained. He clearly thought I was mad, but bade me a cheery farewell anyway before turning back.

Beyond the first headland the Jandía peninsula reappeared, still looking an awful long way away and indeed almost lost from sight in the haze. And yet tomorrow, all being well, I would be camping somewhere on its southern flank. Of more immediate interest was the path, the faint line of which I could now make out, snaking its way along the cliffs into the distance. The way ahead did at least look reassuringly free from steep inclines, but I could see why the guidebook had warned walkers about needing a head for heights. In places the track was effectively on the cliff face, only just wide enough to walk on with loose footing and a near vertical drop to the sea below.

The next hour was immensely enjoyable, indeed one of the finest parts of the walk. Just after saying goodbye to my garrulous friend at Giniginámar, there were three walkers coming the other way. Thereafter I saw no one at all, and indeed no trace of human activity for several kilometres. The scenery was wild and rugged, here and there the path dipped down from the cliff top to cross the mouth of yet another small barranco. But there were no beaches, hence the absence of people, just black rocks covered in barnacles and the top shells that had been used to such good effect in the village. The only sound was made by a lazy sea lapping gently at the rocks. It was ideally cool, there were even a few spots of rain in the air from time to time, I was in no hurry, and perhaps most importantly of all there was no heavy weight on my back. For once I could simply enjoy myself. Sadly, though, my impression that the walk would be fairly flat proved to be false. At the start of the day I would easily have coped with the steeper bits, but now my legs had gone and every incline was hard work.

This stretch of coast was almost devoid of plant and bird alike, although a Yellow-legged Gull or two drifted past now and then, eyeing me with interest, and there were a few Rock Doves nesting on the cliffs. Otherwise, I could almost have been on a planet where life had yet to emerge from the sea.

For some reason, in the complete solitude of this remote coast, I started to think about my father, who had died more than ten years before. He loved walking and although I have no religious convictions, I couldn't help wondering whether he was with me now. He would have approved of my impromptu walking stick, too: he always had one with him when he went walking. And that made me wonder, in a feverish sort of way, whether his spirit might have played a hand in my finding the stick. Now that I thought about it, it seemed an odd coincidence that I had found a stick of the perfect length lying about in the middle of nowhere, and that it had proved to be so indispensable on today's walk.

My father never went abroad. He always said it was because there was too much to see in Britain to bother doing so. Whether this was the real reason or not I don't know, but the truth is he was brought up in a world where foreign travel just wasn't an option for a working -class lad, except as a foot soldier or deck hand perhaps. Anyway, I certainly hoped that he was with me to enjoy this moment in a land that he never saw, but knew something of from all the slides I showed him over the years.

★

An hour from Giniginámar, and about halfway to Tarajalejo, I crossed a large barranco and was snapped out of my day dreams by the appearance of goats and a pile of dumped clothes. There was a beach of sorts at the mouth of the barranco, and although it was remote, there were people on it. They had parked their yellow pick-up truck at the end of a rough track high above, and were camping by the cove. Unspoilt tranquillity was over and I was coming back into civilisation. A few hundred metres further on was the more extensive pebble beach of Playa de Caracol, which can be reached by vehicles along another rough track, and it was a veritable camp of caravans and tents. The change in scenery was accompanied by an immediate increase in birdlife. A pair of Ravens watched me warily from a rock, there were Fuerteventura Chats in the barranco, and the first Berthelot's Pipit for hours called cheerily nearby. It seemed I was back in the real world.

The last two kilometres of the walk to Tarajalejo were largely uneventful: the sense of exploration and reverie had evaporated and it was simply a matter of completing the walk. However, there was to be one last surprise in the tiny little cove before the town. Here, at the end of a steep track, was the simple hut of a modern-day hermit, carefully made from palm leaves, wood and plastic sheeting. Above it a small black flag fluttered in the breeze, and on the hillside close by was a square 'mosaic' of black stones, perhaps three metres across, on which had been painted a white cross. Below it was a long plank on which the words 'Gloria a Dios' (glory to God) were written. Suddenly, as I was contemplating all this, a large, sand-coloured dog, which must have been sound asleep when I walked past the hut a few minutes before, came flying out towards me, barking fearsomely. I fingered the dog-dazer but the hermit called him back, and I carried on to the end of my walk at Tarajalejo.

Box 19 The Ginny Plant

Calotropis procera is a member of the milkweed family, and is no native here. I have come to know it as the 'Ginny Plant', for no better reason than that it is mainly found around the old Berber village of Giniginámar. It is a striking plant, 3m high or more when mature, with big downy leaves arranged in opposite pairs up its woody stems, looking rather like stacked saucers. After rain it produces handsome clusters of wine-red flowers tipped with creamy white. It occurs throughout the desert fringe of North Africa, the Middle East and as far east as India. It was introduced to Giniginámar as recently as 1967, so was something of a newcomer when we found it growing in a valley near there in 1979.

The 'Ginny Plant' has many names within its wide area of distribution, amongst which are Sodom Apple and Giant Milkweed. The white latex is cardio-toxic, and in the past was used as an arrow and spear tip poison. In India, all kinds of medical benefits are attributed to its latex, many of which are at best dubious, but all are agreed on one effect: it is strongly purgative.

The plant spreads readily, both by wind dispersal of its seeds and through birds eating them and depositing them elsewhere in its droppings, and is a characteristic weed of dry wastelands in its native area. In Fuerteventura it is spreading but remains scarce, being found only where there is more moisture, such as in the base of barrancos. Its spread along the south coast of Fuerteventura is probably restrained here by the severe aridity, but if it gets into the wetter hills it might just become a troublesome weed. On the positive side, though, another of its established properties may prove beneficial: it improves soil fertility.

Twenty-four

Tarajalejo and Easter Sunday

Tarajalejo was apparently one of the more important harbours on the island in the seventeenth century, though there is nothing to suggest this today. It is a town of two halves: an eastern residential area that has grown up around the old fishing village, and a tourist area to the west beyond a wide barranco. Tourism started here quite early on, and it was already a rather exclusive German resort in the late 1970s. There were two notable things about Tarajalejo in those days. Firstly, this was where the tarmac ran out on the way to Jandía. Secondly, there were strange mushroom-like parasols on the beach, constructed from wooden posts and palm leaves. However, as at Giniginámar, the lack of golden sand means that the scale of development has been relatively small.

I stayed at the four-star Bahia Playa hotel, which had recently been rebuilt. I must admit that I was impressed: it was surprisingly cheap, it had the best rooms you could wish for, and the staff were utterly charming. As I dragged myself in to register at the end of a long, tiring walk, dishevelled and with a walking stick in my hand, the attendant must have wondered what on earth it was that stood before him. But he was determined to do his best for me, and the free glass of champagne I was given was more than welcome. The room had a huge bed, a very welcome bath and a balcony overlooking the landscaped garden and pool. More unusually, one wall was dominated by a huge photograph of goats grazing on a plain somewhere in the north of the island, which made me feel quite at home.

★

At dawn the next day I was woken by the unmistakable fluty calls of a Bulbul. Bulbuls are mainly tropical birds, and many are popular cage birds: they have rich voices and are easy to keep. The Common Bulbul, a drab bird about the size of a Song Thrush, is a common bird in gardens and thickets only 100 kilometres away in Morocco, but does not occur in the Canary Islands. Bulbuls seem to be unable to cross even short stretches of ocean.

Although I have never seen an escaped Common Bulbul in Fuerteventura, I have come across escaped Red-vented Bulbuls in Corralejo several times in recent years. They are native to India, and rather more exotic-looking than their Moroccan cousins: generally grey but with a black head and breast, and the bright red vent that gives them their name. I wanted to find out whether the bird calling outside the hotel was this species so I hurriedly got dressed, grabbed my binoculars and dashed out. It was indeed a Red-vented Bulbul (Plate 35). Corralejo and Tarajalejo are so far apart (I knew because I had just walked it!) and there is so much unsuitable habitat for Bulbuls in between, that they must have been the result of separate escape bids or deliberate introductions. The extent of suitable habitat on the island is probably too small for them ever to establish a viable wild population, but time will tell.

★

It was already uncomfortably warm by the time I had eaten a hearty breakfast and checked out. After yesterday's walk without the heavy pack, the weight on my shoulders seemed heavier than ever and I was struggling even before I got out of Tarajalejo. It was going to be a hard slog, and I knew that the stretch of coast in front of me was not particularly attractive. The walking stick, which had proved so useful on the rocky slopes yesterday, seemed superfluous now that I was on flatter ground, in fact more of a hindrance than a help. However, I had become attached to it and decided to keep it with me, at least for the time being.

What used to be the main road to Jandía is now a traffic-free promenade, and I was amused to see that the old wooden parasols are still there. In the first little cove beyond Tarajalejo the low tide had exposed a platform of black rock, where several people were collecting shellfish and loading them into vans. After an hour or so I found myself on a path that climbed higher above the sea. From here I could make out the southern point of Jandía far in the distance, where I was booked into a hotel the night after next. It seemed an awfully long way away. Then, suddenly, there was the village of La Lajita in the valley below. Beyond it I could see the great swathe of palms and other trees of the aptly named Oasis Park. The Oasis is home to the island's zoo, and I was planning to stop there later for refreshments.

As I made my way through the sleepy village streets, church bells suddenly started as if to welcome me, and it dawned on me that it was Easter Sunday. To the side of the plaza, and right on the shore was the simple church with its splendid setting. A row of stately palms is all that separates the church from the beach, where brightly coloured fishing boats were hauled out on the shingle (Plate 21). The belfry's single bell was being rung with great gusto.

I suddenly remembered the Mars Easter egg that my wife had given me. I found it just as I was about to give it up as lost, tucked away in one of the rucksack's forgotten recesses. How remarkable that it had made it this far in one piece! It reminded me of my first trip abroad, one cold Easter thirty years before, when I had gone to the south of France with two friends. We spent one glorious week at a farmhouse in the hills above Montpellier, enthralled by our first exposure to the exotic sights, sounds and smells of the Mediterranean garrigue, then a week camping around the Camargue, hauling heavy rucksacks in much the same way as I was doing now. We were penniless students, and shared a small egg on Eastern Sunday amongst the Roman ruins of Arles.

Eating the egg with due ceremony, I suddenly felt very alone. The connection with family and home had been made and the solitude of my journey was brought into sharp relief. Then I made a decision. I really didn't need the walking stick any more; it was simply an extra burden now, so it made sense to leave it here. Unfortunately, though, it wasn't that simple. I had noticed before during the walk that I somehow became attached to everything I had with me and could hardly bear to part with anything. Even a plastic water bottle that had been used for a few days was hard to part with: in the absence of human company, everything I carried was a part of the journey, almost a companion. The walking stick, which had stimulated thoughts about my father, seemed now to be an inseparable part of me. Irrational as it may seem, it was almost as though I was saying goodbye to him. I leant it carefully against one of the palm trees, the sound of Easter hymns drifting out through the open door of the church. I said goodbye to Dad's stick, wished him peace, shouldered my rucksack and began walking again. There were tears streaming down my face as I walked up through the village, and I had to fight the irrational urge to turn back and retrieve it. The walk was clearly getting to me.

★

There is a roundabout on the main coast road at the turning to La Lajita village, and on it is my favourite of all the sculptures on the island. The sight of it soothed me into a more reasonable state of mind. It is about 15m high, and consists of a dull green, sparingly branched, spiky structure. Passing tourists might think it was an abstract design, but, in fact, it is a lifelike representation of a sprig of the *Launaea* bush, one of the most characteristic plants of the island. The likeness is so good that, for me at least, it evokes the plant's distinctive goaty smell.

Now that I was near the Oasis Park and its zoo, I noticed several exotic birds flying about, including a Myna bird and a Ring-necked Parakeet, which were evidently free to fly where they pleased. It was as though I had suddenly been transported to India! Perhaps this was the origin of the Bulbul I had seen earlier in Tarajalejo.[†]

The zoo is a wonderful place for a family outing. I had taken my daughters there one scorching hot August day six years earlier, and they had loved it, although my younger daughter Katy was fearful of the alligators. The lush tropical gardens with their dense shade are a

† There is now a small feral population of this bird breeding in the La Lajita area.

very welcome change from the usual barren landscapes, and there is a good selection of well-kept animals. But, this being a Spanish zoo, it is no surprise that the star attraction is the parrot show: the Spanish simply adore parrots.

As an ornithologist, I must admit to having mixed feelings about parrot shows. I have been lucky enough to see some of the more exotic parrots in their natural habitats. In Ghana, I was thrilled to see Grey Parrots flying through misty rainforest; in Goa I could have wished for nothing better than beautifully plumaged Blue-winged Parakeets to mark my fortieth birthday. In Australia, which is truly the land of parrots (there are more than fifty species), I witnessed the ear-piercing screeches of Sulphur-crested Cockatoos over the Queensland rainforests, the even more incredible oral barrage of a nesting colony of hundreds of Little Corellas (their smaller cousins) in gum trees along the Murray River, and flocks of the gorgeous Pink Cockatoos in the outback. Best of all, I was very lucky indeed to see a Ground Parrot, one of the rarest and most difficult of all parrots to see.

So I come to parrot shows as something of a sceptic, not entirely comfortable with the concept of teaching them to mimic human behaviour: wearing sunglasses, riding a bicycle, relaxing in a deckchair, roller skating, and so on. And yet it is an extraordinary show.

Now, as I sat in the café sipping coffee, I could hear the parrot show in the distance, and some of the parrots that take part in the show were in a cage nearby. I recognised them as the same parrots we had seen before. Parrots are generally long-lived creatures, so it would be possible to repeat essentially the same show every day for decades! You do wonder about the sanity of the people who run the show though. One suspects they get through presenters much faster than they do parrots!

After resting at the café for an hour or so, it was time to head on. I made my way past the line of camels waiting for the next batch of tourists, and out into the heat. It would be another eight kilometres of dull walking to the resort of Costa Calma, and the promise of a hearty lunch.

★

The busy coast road was no place to walk, so I headed down towards the sea. There was no obvious path to follow: it was simply a matter of picking a way west. After more than an hour of trudging across one of the dullest bits of the island I found myself at a rocky beach with a big sign proclaiming that there was 'NO CAMPING'. Despite this, there were lots of caravans and tents, and the beach was full with happy people. I took my rucksack off and stopped for a rest. I removed my socks and shoes and gratefully paddled my blistered feet in the cool sea. I thought that the salt water would do them good. Unfortunately, though, the beach's black grit got into the open blister and wounds, and I had to get it out as best I could before setting off again.

Further on, Matas Blancas came into view. Thirty years ago, it was an insignificant cluster of old fishermen's houses and an old stone tower on a small bay. We had camped here one night in 1979, during the Fuerteventura Houbara Expedition. It was a remote spot then, ten kilometres down the dirt-track from Tarajalejo. Now it is just a few hundred yards from the fanciful architecture of the Costa Calma resort hotels.

Perhaps the old stone tower and some of the dwellings that surround it were constructed from the pre-Conquest wall that once separated Jandía from the main part of Fuerteventura. The southern end of it must have been hereabouts. On a whim, despite the weight of my pack, I headed inland to see whether I could find any trace of it. I crossed the busy road and rested for a while in a small tamarisk thicket a few hundred yards up the barranco, where I thought I might find a few birds. At first it seemed completely lifeless, but then a Robin called and I watched it flicking furtively through the deep shade. Robins are scarce but regular winter visitors to the island.

On the plains beyond I found no trace of the old wall. There are places nearer to La Pared where it is said still to exist, and indeed there is a wall marked on my 1971 military map, about a kilometre further north than I went, which is presumably it. If I had bothered to dig the map out of my rucksack, I might have walked the extra distance. Then again, I was tired and very footsore at this stage, so perhaps it is a good thing I didn't.

★

The sand-covered isthmus north of Costa Calma is one of the most important habitats on the island. As with other sandy areas such as Lajares and Corralejo, the dominant plants include the Yellow-flowered Restharrow *Ononis hesperia* and the little grey succulent sub-shrub *Polycarpaea nivea*. A much rarer plant that also grows here is the endemic Medusa's-head Bindweed *Convolvulus caput-medusae*. Although there is a small population in Gran Canaria, the largest colonies, totalling an estimated 50,000 plants, occur on Fuerteventura. There are none in Lanzarote or the nearby islands. It is an unusual bindweed in that it is a spiny, cushion-forming shrub up to about half a metre high and a metre across.

The isthmus is also important for birds. This is one of the best places on the island for Houbara Bustards and Cream-coloured Coursers, and by far the most important habitat for Black-bellied Sandgrouse, which can be seen here in good numbers.

Today, though, I was more interested in getting to Costa Calma for a much needed lunch and a rest from the sun and the endless walking, so I pressed on, crossing the La Pared road and made my way into the resort. On my left was the most fanciful roundabout sculpture on the island, similar to some of those to be seen on Lanzarote. It is a series of steel spheres and discs designed to move round in three dimensions as it catches the wind. It is actually quite interesting to watch, exactly the eye-catching kind of thing you don't want on a roundabout in fact!

★

It was well into the afternoon by the time I finally reached the cool shade of the plantation that separates Costa Calma from the main coast road. When the winds drift migrant birds across the sea from Africa, this plantation, a veritable forest on this barren island, can be full of small birds. Today I had no energy to look. The pack now seemed like a lead weight and my feet were shot to pieces. I badly needed rest and food.

There were few places open but I eventually found somewhere to eat. In the absence of company I propped my rucksack in the seat opposite me. With my hat on top it looked like a person sitting there, but a mute one of course. I was starting to feel homesick and, for now at least, any sense of excitement had gone. At least the food was good though.

I often noticed that when I stopped walking I started to feel lonely. On this occasion the feeling was particularly strong. I was overcome by the urge to finish the walk, and I made up my mind there and then that tomorrow would be the last day of walking. I would finish at the lighthouse at the southern tip of the island at Jandía Playa. I had no stomach for the hard walk to the western tip. Having made this decision, I headed off in search of enough provisions to get me through the last twenty kilometres. It was cooler now and I wanted to get as far as I could in the remaining daylight. Unfortunately, I had forgotten that it was Easter Sunday, so almost everything was shut. At a garage I managed to buy three litres of water and some bread, a small bar of chocolate and a small carton of fruit juice. Together with my remaining dried fruit, an apple and a few chocolate biscuits, this would have to be enough for the final push.

TWENTY-FIVE

Sand

It was sand all the way now: one continuous beach to the Morro Jable lighthouse at the southernmost point of the island. The tide was ebbing and the wet sand was perfect for walking: firm but slightly giving. I strode on, determined to make as much progress as possible before nightfall. Sun-worshipers heading back into Costa Calma pretended not to notice me plodding along with my rucksack, but when I stopped a couple of middle-aged Germans and asked them to take my photograph they were very happy to oblige. Behind me the sun was already low over the mountains of the south coast, and there were showers of rain sweeping across them. I could see Tarajalejo looking very distant now, and beyond that the sun glinting on the masts above Gran Tarajal. I was filled with a sense of real achievement.

Two kilometres west of Costa Calma the beach starts to widen out. There is a sandbar offshore defining a shallow tidal lagoon, or more precisely a vast expanse of sand flats covered only by the higher tides. In its natural state most of it would be covered with a dense growth of saltmarsh. However, the huge expanse of sand is very popular with holiday makers staying at the nearby Los Goriones Hotel, which is right by the beach overlooking the lagoon, nestling amongst palm trees and exotic shrubs. The marsh is much degraded through trampling, but is still dotted with dark green saltwort *Arthrocnemum macrostachyum*, a characteristic plant of saltmarshes on the island with leafy branches like ostrich feathers.

Despite the people, the lagoon is still attractive to water birds. As I walked past it there were two Cattle Egrets (a new colonist to the island), Little Egrets, a Spoonbill, and a few smaller waders. Early in the morning, before the walkers have spread out from the hotel, the sand bar beyond the lagoon is an important resting place for terns and other shorebirds. There can be hundreds of Common and Sandwich Terns in winter and especially during the migration season. Rare Slender-billed and Audouin's Gulls have also been seen here in recent years. Where they all go to during the day remains a mystery.

The sun was very low by the time I passed the hotel, and so too were my energy levels. I wanted to stop but I needed to find somewhere quieter to pitch the tent, so I kept walking slowly onwards for another three quarters of an hour until I found a suitable spot. There were still a few stray walkers and joggers around, but I decided to make camp anyway on a slight rise in the dunes, overlooking the sea. It was to be my last, and certainly the most beautiful, camp site of the walk (Plate 36).

Once I had pitched the tent I sat outside in the failing light to eat my rations. In front of me the empty lagoon was wet and glassy, reflecting the clouds, and there was the sound of the surf crashing on the distant sand bar. I reflected on the day that was now drawing rapidly to an end. It had certainly had its ups and downs, but it had not been as hard as I had feared at the start. The effort had been worthwhile, because now I had just twelve kilometres to cover.

As darkness fell I clambered into the tent. I was ready for sleep and keen to get up early to complete as much of the walk as possible whilst it was still cool. Inevitably, a Stone-curlew called somewhere nearby and before midnight I was half woken by the sound of a Sandwich Tern flying low over the tent. Then came the sound of fireworks in the distance, seemingly from La Pared, which set the Stone-curlew off again. At four in the morning I woke yet again, this time to the thrilling sound of hundreds of terns and waders out on the moonlit sandbar, their excited calls rising and falling in excitement in response perhaps to the motion of the waves. The moon was flooding the tent with light.

On a whim, I decided not to turn over but to embrace the moment and start walking along the beach in the moonlight. What better way to end the walk? I broke camp as quickly as I could and began walking. Disappointingly, the moon disappeared behind the clouds almost immediately, but the noise from the sandbar was undiminished and, amazingly, a Lesser Short-toed Lark was singing high in the ink black sky.

Without the moonlight it was more difficult to walk along the beach than I had hoped. I got out my head torch, although there was just enough light to follow the edge of the beach most of the time. Looking back I could see the slow flash of the Entallada lighthouse, now about thirty-five kilometres away.

Now and then there were half-seen buildings a little way from the shore, and I was wary as ever of dogs. Had there been any they would surely have been a real problem for a stranger walking past in the night, but thankfully all the buildings were uninhabited. Occasionally there were lights from the coast road, but for most of the way it was very dark and utterly peaceful. I kept walking on the deserted beach, enveloped in the cool darkness of the night, accompanied by the gentle sound of waves surging over the sand.

The first light of dawn came just before six. I kept walking as the light over the sea slowly increased and colours spread into the clouds. Half an hour later I came to a deserted beach-side café with rows of sun loungers spread along the shore. The furthest was perfectly placed for a rest and a bit of breakfast. A cup of coffee would have gone down a treat, but you can't have everything.

The night ended for me just before seven, when a Fuerteventura Chat sang somewhere by the beach, and there were suddenly five dinghies out to sea. Then the sun came up above the clouds and slowly but surely people began to appear, some out for an early morning jog, but others walking purposefully, laden with bags and keen to claim the best spots out of the wind. In common with other beaches on the island, there were quite a few roughly constructed stone shelters in the dunes, but also some very fine ones. The best were so well crafted as to look almost like bronze-age huts. They were skilfully built by someone who knew all there was to know about dry-stone walling. Their makers must have been very proud of them, and no doubt rather possessive. Some, indeed, had towels in them that must have been there all night (all week?), presumably to deter even the earliest risers from trespassing into their 'territory'.

This prompted me to explore a fantasy. In a few centuries from now, what would archaeologists make if they unearthed these fine structures from the accumulated sand? They might search in vain for remnants of a roof, but in the centre would be the partly decomposed remains of a beach towel, weighed down at the corners with stones.

I walked on, passing some of the more exclusive hotels on the rocky ground overlooking the beach. I was ever hopeful of spotting a café where I could get breakfast and, more importantly, get the pack off my back for a while. But I never did find one that was open for business, so I kept plodding slowly along the interminable beach for another two hours until, rounding a headland, the lighthouse and the end of my walk came into view. After almost another hour of soft sand, I finally reached the lighthouse. I had made it to my journey's end, or at least one possible end. I felt a need to touch the lighthouse in a ceremonial kind of way.

TWENTY-SIX

Jandía Resort and Morro Jable

When I strolled into an English sports bar on the main street in the Jandía resort in search of breakfast, I had been walking for five hours. I was a little surprised, therefore, to find that I was too early! Disappointingly, the coffee was instant, though nevertheless very welcome. There was a television on in the corner of the bar. Back home in England the weather was bitterly cold, and the racing at Fakenham, only fifty miles from home, had been cancelled because of snow.

After breakfast I made my way wearily along the street to the rather grand-sounding Dunas Stella Jandía Resort Aparthotel, where I had booked a three-night stay. This is an old resort by Fuerteventuran standards, and is set in a sub-tropical paradise surrounded by tall palm trees. Before the road to the resort was completed, it was at the end of the long desert track from the main part of the island, and effectively in the middle of nowhere. It was untouchable luxury for me then, living as I was on the bare minimum, just as much as it was for the islanders. Now I had the means to sample it, and I was looking forward to some luxury.

The 'resort' is a large complex of buildings spread over many acres, set on a fairly steep hillside. At the entrance from the main coast road there are manned gates, from which a splendid palm-lined boulevard leads up through the complex with exotic gardens on either side. Just inside the gates on the right is the resort's own small zoo with waterfowl, monkeys and a crocodile show!

I was quite exhausted, so it was disappointing to find that I had a long, steep walk up through the palm trees to the main reception. The reception hall was, however, very grand indeed, a great airy space with marble everywhere and fountains playing into a large pool. It was busy with people arriving with luggage, collecting keys and debating with the receptionists. However, my room wasn't yet ready, so I made use of one of the many luxurious chairs and fell asleep to the soothing sound of running water.

★

When my room did become available, I was pleased to find that the southernmost point of the island and the lighthouse that marked the end of my walk were in full view. At night, the lighthouse's double flash lit the room through thin curtains. I spent the rest of the day relaxing in the cool air, resting my feet, and letting my mind wander over what I had achieved and what would happen next.

I had reached an end point, but was it to be the end point? My intention at the outset had been to walk to the very end of the island, which was still twenty-three kilometres to the west. I had almost given up completely on the second day of the walk because of blisters and the dangerous climb east of La Oliva. My nerve had nearly gone then, but it would have been a big mistake to have given up and, despite the difficulties, there had been some wonderful moments. The highlights that came quickly to mind were the immense sense of freedom and achievement I experienced on the 'pilgrim's path' in the Betancuria mountains, the thrilling sound of distant terns on the sandbar under the full moon last night, and the lights coming on far below from my restaurant 'camp' site on Morro de Velosa.

There had undoubtedly been bad times, too. The last few kilometres of the walk to Fustes were really hard, physically harder than anything I had ever done before. I had run out of water, my feet were terribly sore and my energy had gone. If I had not reached Fustes when I did, I would very soon have been badly dehydrated and in a serious situation. But the bus shelter at Teguital stood out as the absolute low point of the walk: the closed restaurant, the farce of the police check-point and the painful, never-ending walk from there to Las Playitas. And yet on both occasions there was the indulgence of a comfortable hotel at the end.

And then there was the chance finding of my Nicotiana walking stick, just the perfect size and very light, that served me so well on the glorious walk from Las Playitas to Tarrajalejo, and the emotional turmoil it somehow catalysed at La Lajita.

★

The following day I took a gentle stroll down the coast to the town of Morro Jable. The coast road is an ugly affair, lined with tourist shops and bars, but now there is a new landscaped promenade linking the resort area with Morro Jable. There were Monarch butterflies flitting though its palm trees and exotic shrubs, and glorious views over the golden sands below.

About halfway between the resort and Morro Jable, there was a faint, squeaky bird call coming from the dense foliage of a tall tree. It sounded suspiciously like a Yellow-browed Warbler, a species that has only been recorded a few times in the Canaries. Sure enough, after a few minutes it hopped into view and I could see that it was indeed this species. The Yellow-browed Warbler is tiny, only just larger than a Goldcrest, which is Europe's smallest bird. As its name suggests, it has a long, yellowish strip over the eye and, unlike most other warblers, a bold wing pattern. Although it nests in the vast taiga forests of Siberia and rhododendron thickets of the Himalayas, and normally winters in the tropical woodlands of south-east Asia and eastern India, it has somehow become a scarce but regular migrant to Western Europe over the last few decades. This being the case, they must winter in increasing numbers in south-west Europe and north-west Africa, and this bird was surely wintering here in Fuerteventura.

★

Morro Jable itself is a working town with plenty of cafés and other facilities. Close to the old harbour on a narrow passage is the Pub El Barco, a section of an old steamship, complete with portholes and lifeboats. To reach the modern harbour, I walked through the town, and followed a pleasant path down through a landscaped urban space where African Grass Blue butterflies flitted over the lawns. The unattractive concrete harbour is now the stopping point for ferries to Gran Canaria and beyond. I walked to the end of the harbour arm, to mirror the very start of the walk in Corralejo. There was no sense of triumph in doing so, the question of whether to carry on walking to the end of the peninsular remaining unresolved.

★

When planning the walk, I imagined that when I got to the Jandía resort I would walk to the top of Pico de la Zarza (Box 20), which is the highest point on the island. Now that I had arrived though, my feet were too sore, and I needed to rest in case I decided to walk to the far south-west point. That being the case, I set myself a much more leisurely task for the day: to undertake a thorough search of the Stella Canaris grounds for migrant birds.

To that end I spent a whole morning peering into thickets of exotic vegetation, walking slowly in the pleasant shade of palm trees with binoculars at the ready, and generally investigating every nook and cranny of the complex. As a result I recorded just sixteen species of birds, and a strange assortment they were. In the 1950s, before the tourist complex had been built, only five of them would certainly have been here: Spanish Sparrow, Linnet, Spectacled Warbler, African Blue Tit and Little Egret. A further two, Song Thrush and Blackcap, which are winter visitors from Europe to the lusher spots on the island, might have been here if there were any trees and shrubs on the site. Collared Doves, which are so common now, were unknown in Fuerteventura until the 1990s, and Laughing Doves have been seen only in the last few years. Both seem to have colonised naturally from Africa, although the Laughing Dove remains very scarce, and it is not yet clear whether the species will become permanently established. Another recent arrival is the Cattle Egret, a group of which was feeding in the saltmarshes in front of the hotel throughout my stay.

The avifauna of the island is constantly changing, and seems generally to be getting richer.

All the other species were exotic free-flying birds from the zoo. Strutting around the manicured lawns were a dozen Peacocks, two splendid Crowned Cranes, and a couple of glossy black Hadada Ibises. Last but not least were the parrots, which kept up a constant screeching in the palms by the zoo. Ring-necked and Monk Parakeets have nested in the dense stand of trees at the zoo for many years, and the Monk Parakeets, at least, seem to be fairly well established in the area.

★

The Dunas Stella Jandía Resort is certainly an unusual place. It has something of the air of an old colonial palace about it. But my ornithological ramblings took me to parts of the complex that were closed down and decaying. One of the incidental benefits of being a birdwatcher is that you get to see things that other tourists simply don't. I can think of no other hobby, except perhaps an insatiable curiosity bordering on nosiness that would result in such an intimate knowledge of the local surroundings.

All around the complex there are marble staircases with peeling paintwork and cracked slabs, many of which now lead to apartment blocks that are unused for lack of repair and investment, presumably because the resort no longer attracts as many visitors. There are once grand staircases that are so overgrown that it would now be impossible to use them. And there are cats everywhere. They must live in the abandoned apartment blocks, many of which have smashed windows and broken doors. They seem almost to own the place, and are kept under control only by the lordly Crowned Cranes, who are not susceptible to their feline charms.

★

Box 20   Pico de la Zarza (Plate 22)

Pico de la Zarza is the highest point on the island at 812 metre. There is a good track to the summit from the Jandía resort, which is described in the Sunflower walking guide. On one occasion, I also walked to the top from the ominously named Barranco de Mal Nombre (barranco with the bad name). This proved to be a much more difficult approach, but included an exhilarating walk along the cliff-top in swirling mist.

Walking up from the Jandía resort, the gentle contours of the southern slope give no warning of the extraordinary views from the top. Suddenly, the mountain comes to an end: it is just a sheer drop hundreds of metres to the narrow coastal plain below. The views along the escarpment of the peninsular are really spectacular, but an even greater lure for me are the rare plants that grow on that inaccessible cliff. Uniquely for this island, the cliff-ledges are bright green all year round. This is because of the persistent mists that form around the upper part of the cliff, where the northerly winds are forced violently upwards by the mountain barrier. Some of the plants that grow here are found nowhere else in the world, such as the daisy-flowered shrub *Argyranthemum winteri*, and the blue-flowered *Echium handiense*. Also here are the last surviving remnants of laurel forest on the island: a few struggling shrubs clinging to bare rock on inaccessible ledges that not even goats can reach. With extreme care, it is possible to access some of the higher ledges from the top, where at least some of the rarer plants can be found.

As recently as 1999 a new species of plant was described from the summit of Pico de la Zarza. Admittedly, it was a rather inconspicuous moss – *Orthotrichum handiense* – growing on the stems of the shrub *Asteriscus sericeus*, which is abundant here. Another recently described species from the Jandía area is the Jandía Clearwing, a wasp-like moth that appears to be restricted to Fuerteventura.

The remote Cofete plain that lies at the bottom of this escarpment is the site of an ongoing Loggerhead Turtle reintroduction programme, which started in 2007 when eggs were buried in the sand. Good numbers of young turtles subsequently made their way out to sea. The plan is to continue to introduce eggs in this way for ten years in the hope that enough adults will make it back there to establish a viable population. Apparently, there is evidence that Loggerhead Turtles nested in Fuerteventura until they were exterminated for food 300 years ago.

The name of this mountain is an interesting one, as *zarza* means blackberry. There is indeed a species of blackberry, *Rubus bollei* growing on the mountain. Today, it is restricted to inaccessible ledges on the cliffs below the peak. However, for the mountain to be named after it, it must surely have covered the upper part of the mountain when it was first climbed by European settlers.

TWENTY-SEVEN

A decision is made

At the end of my last day at the Jandía Resort, I made the decision to walk to the end of the peninsula. I phoned my wife and explained my change of plans. She was rather surprised as I had said quite adamantly only a few days before that I couldn't see the point in doing it just for the sake of it, and I wasn't going to bother.

I had come up with a plan that would make the walk to the end of the peninsula possible, and it involved a car. The idea was to hire a car and drive down the dreadful dirt track to the end of Jandía, and try to hitch a lift back to Morro Jable. Then I could complete the walk knowing that there was a car waiting for me at the far end.

I booked a car for the following morning and turned in early. When I woke up the following morning, though, I was still uncertain. I really didn't fancy the long, hot walk down the busy dirt track with its dust and dreariness. And yet the nagging feeling that I *had* to walk to the point would not quite go away. The ultimate question, it seemed to me, was how I would feel when I got home if I hadn't completed the last section.

I went through 'yes' and 'no' a number of times over breakfast, and decided that it would come down to what felt right when I picked up the car. When I did pick up the car, three factors made the decision for me. First, there was a prominent sign displayed by the desk explicitly forbidding hire cars from being driven off tarmac. I knew that the car would not be insured on the dirt track, and had known that on previous occasions, but this did seem rather pointed. Secondly, the car turned out to be a brand new and very pretty looking gold Corsa. I would certainly be a bit self-conscious driving down the peninsula in it, and I wasn't sure I had the heart to submit it to fifty kilometres of rough dirt track. Finally, and this I think was what really swayed it, it felt good to be in possession of a car and with the freedom of the island again. My decision was made at last. There were better ways of spending my last few days on the island than struggling down a dirt track with a heavy pack for the sake of it. I got in the car and headed east away from Jandía.

★

The morning did not go well, and by the end of it I was thoroughly morose. My first stop was at Los Goriones Hotel. I wanted to walk out onto the sandbar where I had heard hundreds of birds calling in the moonlight on my last night under canvas. But I had left it too late in the day, and after half an hour of hot walking I realised there were no terns or waders to be seen. As usual they had melted away after the first few beach walkers had appeared.

Driving on past Costa Calma and La Lajita zoo, I felt a sadness creeping into me. I was reminded of the loneliness of this part of the walk and the hard effort that had been involved on what was to be the penultimate day of the walk. I stopped by the little church on the beach in La Lajita to retrieve the stick I had left propped against a palm tree on Easter Sunday. I wanted to take it home as a souvenir. But it had gone! At first I couldn't believe it, and searched under all the palms before accepting that it simply wasn't there any more. I walked up and down the beach in case it had been picked up and carelessly thrown aside. But no, somebody had taken it. I comforted myself with the thought that an old man might have found it and had recognised its value as a walking stick, but the reality is that it was probably tossed into the sea by a child.

At least passing Monte Cardón was a good feeling. I could look at its top and know that I had been there: I had conquered its steep rock. But the melancholy would not go away. Next, I stopped to take a photograph of the bus shelter at Teguital, the absolute low-point of the trip. I don't know why, but I was surprised to find that the rubbish I had put in the bin was still there. Perhaps it was because so much had happened to me in the meantime. Now at least I was free to drive on.

As expected, the sheep I had left dying beside the track was dead. It never moved again, probably breathing its last as I was struggling to reach sanctuary at Las Playitas. It looked the same except for a small hole in its flank where a Raven or perhaps an Egyptian Vulture had opened the carcase. Inside the hole was a seething mass of maggots.

Now I turned and headed towards the lagoon at Catalina Garcia, where hopefully there would be some new birds to lift my spirits. I ate lunch overlooking the lagoon. The birds were, in fact, much the same as before, but by chance Derek Bradbury turned up. It was good to have company for a while, and I felt much better by the time I drove on.

My next stop was at Ajuy on the west coast, where a walk to the sea caves and the sea arch at Peña Horadada restored my spirits completely. Then I dove back through Pájara and took the mountain road to Vega de Río Palmas, where I stopped briefly to admire the church with its attractively decorated front and its totemic image of the Virgin.

Lastly, before completing my journey north to Corralejo, I visited the café at Mirador de Morro Velosa, high on the northern rim of the Betancuria mountains, where I had slept out under a full moon on the way south. It was late in the day and, apart from a cheerful waiter, I had the place to myself. I sat drinking coffee alone by the huge windows with their immense view, looking out over the landscape that I had walked through in the first few days, over the mountains where I had realised that my dream to walk the island was within my capability. Now it was done and I was filled at last with an immense sense of achievement.

★

For the last three nights I stayed at the four-star Blue Bay Hotel in a quiet part of Corralejo. Not cheap, but an excellent choice:

friendly and helpful staff, a superb room and great breakfasts in very comfortable surroundings. But I was in limbo, having completed my walk but not being able to go home.

On the first morning, as I stood looking out over the town from my balcony, a Barbary Falcon shot past. Later on, after an early breakfast I went for a leisurely walk over the sandy plains to the south, that are part of the Corralejo Dunes Natural Park (which was described in Box 6). In the past, this was one place where Houbaras were common, and I was hoping to find one now. Unusually there was no wind at all, and in the still air larks were singing everywhere. Eventually, after scanning for more than an hour, I spotted a male away in the distance. I watched him for some time, hoping to see him display, but although he half-raised his neck feathers, he never did. It was probably too late in the season. It was good to know that they are still here, despite all the development that is going on around them.

I headed back into town and returned to the Brisamar café where my walk had started. I sat in exactly the same seat as before, but this time the café was completely empty. I remembered that first morning, the struggle to get everything in the rucksack, and my chat with the English couple. There were ghosts here. I could feel the emotions I had had at the time, a mixture of fear, determination, and excitement. For some reason I was surprised that the English couple were not there, just as it had seemed impossible that my stick had gone at La Lajita. I could see myself outside the café, struggling to get the heavy pack onto my back, desperately trying to get my left arm through the strap and to look as though I knew what I was doing.

During the last few days, time passed slowly, and I took the opportunity to visit some of the island's museums. One morning I even found myself wandering round the new Las Rotondas shopping centre in Puerto del Rosario. Puerto has sprawled uncontrollably in recent decades, spilling out in all directions onto the desolate coastal plain. There have been some significant changes too. Most importantly, the military area near the centre of town that had been the headquarters of the Spanish Foreign Legion since 1976, when Spain was forced out of what is now Western Sahara, is now deserted. On 22nd February 1981, I had gone to the airport to hire a car and noticed a surprising amount of military activity along the road between the airport and Puerto. Tanks seemed to have taken up significant positions. The following day, the paramilitary Civil Guard stormed the lower house of the Spanish parliament and tried unsuccessfully to install a military dictatorship in Spain. Whether the Legion's military manoeuvres were coincidental or not, I don't know, but there were certainly some unusual activities going on that day. The presence of the Foreign Legion had given Puerto an uneasy feeling, and I suspect there are few who mourned its departure in 1996.

But there is one good reason for visiting Puerto del Rosario: the museum dedicated to the writer Unamuno (Casa Museo Unamuno), which is perhaps the most important museum on the island (Box 21). On the pavement immediately outside the museum is a statue of Unamuno. This is one of the hundred or so street sculptures that have sprung up around Puerto in recent years. They include both modern art and figures of significant local people. There is even one sculpture that is supposed to represent the Fuerteventura Chat, although it bears no resemblance to the bird at all: it is a swan's neck on a pole. I particularly liked the jumble of suitcases, complete with an umbrella and a hat, opposite the port, representing both those Majoreros who emigrated to the American colonies and the present-day settlers from Europe who form an important part of the island's culture today. It reminded me of my first visit to the island all those years ago when I lost my day sack very close to here, and was indeed wearing a rather splendid hat. I also liked the water carrier, a man carrying two large cans, a common sight apparently until the late 1960s.

Opposite the Unamuno museum is the main church in Puerto, Nuestra Señora del Rosario, from which the town now takes its name. To me, its exterior is the most unsatisfactory of any church on the island. The tower is a horrid-looking grey and white affair that looks like it might be part of a cardboard film set. The bell tower was constructed specifically to lodge two bells from Marseille of all places, prior to which it evidently had a small belfry more in keeping with the local style.

A more honest church is to be found in Casillas del Ángel, a few miles to the west. It has a front of black lava blocks, with graceful baroque curves. An antidote to the brutal tower at La Oliva that shows what can be done, even with lava, if the will is there.

★

Box 21 Miguel de Unamuno

Miguel de Unamuno was one of Spain's foremost early twentieth-century writers and intellectuals and, unfortunately for him, he didn't see eye to eye with the military dictator of the time, Primo de Rivera. As a consequence, he was exiled to Fuerteventura in 1924. It is interesting that he was sent to Fuerteventura rather than to one of the distant colonies. No doubt Fuerteventura was sufficiently isolated in those days to make communication of uncomfortable truths difficult, but perhaps exile to Fuerteventura was also seen as a punishment in its own right. If so, this backfired spectacularly. Although he was only on the island for four months, Unamuno became inspired by its harsh landscape and his writings have left an indelible imprint. Fittingly, the museum is in the old Hotel Fuerteventura, where he lived for much of the time, and the island council is to be congratulated in having secured it for this purpose. It is, in part, a museum of early twentieth-century furniture and fitting, including as many of the things Unamuno personally used as could be found, but more important are the extracts from his writings about the island which are displayed on the walls. Through his writing, he conveys the spirit of the island better than anyone else has managed to do.

There is a substantial monument to Unamuno near Tindaya set into the volcanic cone of Montaña Quemada, just off the road to Tetir. It appeared there between my first and second seasons of bustard studies, and was apparently unveiled with much ceremony in November 1980. Today, it is virtually ignored. The real monument to Unamuno is his own writings, and the museum in Puerto is a more fitting tribute.

As with all great writers, you have to read him in his own language to appreciate fully what he is saying, but even when translated into English, his fascination with the harshness of the island is clear in his powerfully evocative writing.

> These denuded, solitary, skeletal hills,
> of this bare island of Fuerteventura!
> This skeleton of land, rocky entrails
> that rose from the depths of the sea, volcanic ruins;
> this reddish frame tormented by thirst!
> And what beauty! It is clear
> for he who knows how to look for the intimate
> secrets of form, the essence of style,
> in the denuded line of skeletal hills;
> for he who knows how to discover a beautiful head in a skull

But if he saw beauty in desolation, as many of us do, he was no heady romantic. Another passage from his writings suggests that he sometimes found the barren hills and its strange plants overwhelming. Of the succulent spurge (*Euphorbia regis-jubae*) that grows so commonly on the island, he wrote:

The acrid, caustic spurge milk
is the juice of the calcinated bones of
the volcanic land that rose from the depths of the sea;
The acrid, caustic spurge milk
is the bone marrow of this thirsty land.
And you have to nourish the spirit
with spurge milk

He was clearly fascinated by form, and made frequent reference to the flora in trying to capture the essence of the island.

All style, even that of nature, is autobiographical. This island of Fuerteventura (fuerteventurosa island!) for example, has style that other islands converted into gardens by man do not have; this island for pilgrims (pilgrims in search of ideas), and not for tourists, this island has style, a skeletal style. Skeletal is the land, these ruins of volcanoes that are its mountains, in the form of camels' humps, the mountains of this camel-like island; skeletal are its camels, that curse its strong frame of hills; skeletal is the aulaga [Launaea], the poor gorse that covers the rocky ground, that shrub that is all spines and flowers, without a single withered leaf, unadorned, lean, bony; skeletal is the tamarisk, this faded tamarind that shakes its mean, lank, grey foliage in the wind; skeletal also is the lump of gofio, of toasted wheat flour, that gofio that is like the skeleton of bread; skeletal are the houses, these houses without tiled roofs, many with bare walls... And all this solemn denuded boniness is autobiographical. With this bareness, Fuerteventura describes itself, it describes itself as it is.

He was wrong about the tourists though, who look the other way and see blue sea and golden sand.

TWENTY-EIGHT

The end

Going home day finally dawned, and now that it had I was sad to be leaving the island. One of the most important events in my life was coming to an end. The walk had been hard, but I could look back on it now with a good deal of satisfaction.

The ferry I planned to catch to Lanzarote didn't leave until the afternoon, so I thought I would spend my last morning exploring the island of Lobos, the volcanic cone offshore from Corralejo. I went to the quay, but for some inexplicable reason the boat to Lobos was just leaving, even though the timetable suggested it should not be doing so for another hour. There were no other boats that would get me to the island and back in the time I had available, so that was that. With nothing much else to do I sat on the quay for a while, and my mind went back to the very start of the walk when two black cats had suddenly appeared as if from nowhere as I prepared myself for the trials ahead of me. At the time I had wondered whether this was a good omen or not. Now that the walk had been completed, it seemed as though they had been.

Having failed to get to Lobos, I found myself wandering rather aimlessly around Corralejo. I thought of visiting the reptilarium but discovered that it is shut on Sundays, so was reduced to buying a *Sunday Express* and retreating to the Rock Café for a pint of Guinness. If you can't beat them join them!

I ended up in another very British bar where I sat watching an

English football match. Oddly, it had also started an hour early. The world was clearly going mad. With the time dripping slowly away I decided to find somewhere where I could at least treat myself to a decent lunch before leaving. I found the perfect place: a restaurant right on the front, opposite the ferry terminus. I propped my rucksack against the wall, ordered a good lunch and settled down to enjoy the last hour or so on the island.

Across the bay cars were being loaded onto a ferry. Slowly but surely it dawned on me that it was the ferry I was supposed to be catching. Then the penny dropped: the world hadn't gone mad at all; it was simply that the clocks had changed. My ferry was leaving in just over five minutes! In one breathless move I shouldered my rucksack, explained to the waitress that I had to catch the ferry, so wouldn't need the lunch after all, and rushed out of the door. I ran full tilt along the promenade despite the weight of the rucksack, desperate to get there before the ferry left. I made it, gasping for air and sweating profusely, and fumbled for my ticket. No use, the smartly dressed, imperturbable young lady at the entrance to the boarding plank explained firmly, I had to get a boarding pass from the ticket office. I pleaded with her to let me on, but it was no use. There was enough time for me to collect the boarding pass she insisted. I ran 200 metres to the ticket counter, waited agonising moments for the required piece of paper, and then ran back to the ship. It was still there, she had been right. I made it with two minutes to spare.

The ferry pulled out exactly on time. Once my heart stopped thumping violently against my chest, and I had regained a certain amount of composure, I glanced down at the quayside. There were no black cats in sight. I settled into a chair to get my breath back and watch the island recede into the distance. I could see the track edging round the volcanic cone of Buyayo, where my walk had started so uncertainly. I tried to remember what it had felt like to be setting out on the walk three weeks ago. And then the engines slowed. In just over a quarter of an hour we were entering the harbour at Playa Blanca on the south coast of Lanzarote. I was still staring at the track, now just a faint, sandy-coloured mark on a distant island.

Postscript

During the six years that passed between completing the walk and publication of this book I returned to Fuerteventura several times. The local council has now established well-marked walking and cycling routes throughout the island and it would probably be possible to walk from one end of the island to the other along these trails, although information on them is rather hard to find. Nevertheless, I am glad to have undertaken the walk before they were created, as some of the adventure would inevitably have been lost had they been in place at the time. Perhaps a few people have already made use of these trails to walk from one end of the island to the other, but surely none will have taken the eccentric mountain route described in this book.

Something new has been learnt during each of these subsequent visits. In April 2013, with my wife and youngest daughter for company, I finally managed to explore the island of Lobos, which sits so invitingly just offshore from Corralejo. What a fine place it is with its unspoilt Euphorbia scrub and coastal lagoons. Perhaps one day the rare Monk Seal will return to raise its pups: there are still a few to be found further south along the African coast. For now, though, we will have to make do with the sculptures of them that have recently been placed outside the visitor interpretation centre.

In March 2012 the Honeyguide holiday to the island, based at the excellent Casa Vieja rural hotel in La Oliva, coincided with a major influx of migrant birds. We saw 90 species during the week, perhaps a record for an island where 50-60 species is the norm. Our most exotic find was a handsome male Bluethroat, a real rarity in the Canary Islands. During the family holiday based at Lajares in 2013 there were also large numbers of migrants, including several dazzling Golden Orioles in the trees around the village, and a Wryneck that took up residence in and around the garden of our villa.

The following September, Howard Taffs and I spent several days scouring the island for birds. It was more than thirty years since Howard had been to Fuerteventura, and so much had changed that it was almost unrecognisable to him. We witnessed a rare late summer thunderstorm, which resulted in an unprecedented arrival of Pied Flycatchers. We counted no less than eighty of these delightful birds in a day. Their flight south to wintering grounds in tropical West Africa had clearly been seriously disrupted by the unusual weather. But the main reason for our trip was to take the ferry to Las Palmas and back in search of seabirds. This confirmed that rare Bulwer's Petrels occur just off the south coast, and indeed are relatively common at this time of year in the seas between Punta de Jandía and Gran Canaria. There were also Basking Sharks, Pilot Whales and Loggerhead Turtles, the last perhaps from the reintroduction programme at Cofete.

Every visit brings more discoveries, and each seems more enjoyable than the last.

Bibliography

Araña, V. and Carraceddo, J. C. 1979 *Los Volcanes de las Islas Canarias / Canarian Volcanoes. II Lanzarote Y Fuerteventura*. Editorial Rueda, Madrid. ISBN 84-7202-014-X

Automobile Association. *Island Map, Fuerteventura*. Island series 6. 1:50,000 scale

Bannerman, D. A. 1922 *The Canary Islands: their history, natural history and scenery*. Gurney and Jackson, London.

Blamey, M. and Grey-Wilson, C. 1993 *Mediterranean Wild Flowers*. HarperCollins. Domino Books.

Bramwell, D. and Bramwell, Z. 1987 *Historia Natural de las Island Canarias: Guía Básica*. Editorial Rueda, Madrid

Bramwell, D. and Bramwell, Z. 2001 *Wild Flowers of the Canary Islands*, Second Edition. Editorial Rueda, Madrid.

Carracedo, J. C. et al. 1980 *Canarias*. Ediciones Anaya, Madrid.

Clarke, T. 2006 *Birds of the Atlantic Islands*. Helm Field Guides. Christopher Helm, London

Clarke, T. and Collins, D 1996 *A Birdwatchers' Guide to The Canary Islands*. Birdwatchers Guides, Prion Ltd.

Clements, W. H. 1999 *Towers of Strength*: Martelo Towers Worldwide. Leo Cooper.

Collar, N. J. 1983 A History of the Houbara in the Canaries. Bustard Studies 1. *Journal of the ICBP Bustard Group*. International Council for Bird Preservation.

Collins, D. R. 1984 *A Study of the Canarian Houbara Bustard (Chlamydotis undulata fuertaventurae) with Special Reference to its Behaviour and Ecology*. M.Phil thesis, University of London

Collins, D. R. 1984 Studies of West Palearctic Birds 187: Canary Islands Chat. *British Birds* 77: pp 467–474.

Collins,D. and Clarke,T. 1996 Birding in the Canary Islands. *Birding World* vol 9 no 6, pp 228–241.

Kunkel, G. 1977 *Las plantas vasculares de Fuerteventura (Islas Canarias), con especial interés de las forajeras. Naturalia Hispanica 8. Ministerio de Agricultura.* Instituto Nacional para la Conservation de la Naturaleza. Madrid

Lorenzo, J. A. (ed.) 2007 *Atlas de las aves nidificantes en el archipiélago canario (1997-2003).* Direccíon General de Conservación de la Naturaleza-Sociedad Española de Ornitología. Madrid

Mercer, J. 1973 *Canary Islands: Fuerteventura.* The Islands Series. David & Charles, Newton Abbot.

Mercer, J. 1976 *Spanish Sahara.* George Allen & Unwin Ltd. London.

Meurer, M. (undated) *Fuerteventura: Magie Einer Insel/Magia de Una Isla.*

Murphy, P. 2005 *Fuerteventura: Spiral Guide.* AA Publishing.

Myhill, H. 1968 *The Canary Islands.* Faber and Faber Ltd, London

Richford, N. 1999 *Landscapes of Fuerteventura, a countryside guide.* Second edition. Sunflower Books

Rodríguez Delgado, O. 2005 *Patrimonio Natural de la isla de Fuerteventura.* Cabildo de Fuerteventura/Gobierno Canarias/ Centro de la Cultura Popular Canaria.

Rodríguez, R. and Paredes, R. 1994 *Fuerteventura. Guías y Mapas Rai. mundo.* Gráficas 7 Revueltas, Seville. Second edition.

Tolman, T. and Lewington, R. 1997 *Butterflies of Britain and Europe.* Collins Field Guide. HarperCollins

Appendix One

Fuerteventura bird list

This list, the order of which follows Tony Clarke's *Birds of the Atlantic Islands* (Helm Field Guides), draws on personal observations during many visits of between one week and four months since 1979, holiday reports submitted to me from various visitors over the last thirty years, published books and papers, and sightings reported in UK birding magazines. The more frequent free-flying introduced birds are included. For information on where to watch birds on the island, see Clarke and Collins (1996).

The status of most migrant birds is quite hard to define because they occur only when weather conditions are exactly right to bring them to the island. Some species can be quite numerous at times, but hardly seen at all in other seasons. A further problem is that many records are from visiting birdwatchers from northern Europe, not residents, so the exact status can be hard to determine.

The list includes 299 species, of which 8 are certainly introduced and either have established feral populations or simply have freeflying populations based at bird parks. This is *not* an official list of the birds of the island, and includes a number of species that have been reported but not necessarily confirmed.

Where I consider there is some remaining doubt about validity of a sighting, I have referred to the occurrence as 'reported'.

A number of other species are seen from time to time as escapes either from private collections or zoos.

Little Grebe	now rare, but in the 1950s there was a thriving colony at Las Penitas, with over 30 birds in 1957.
Black-necked Grebe	a few records
Fea's Petrel	a few records
Bulwers Petrel	Rarely seen, but has nested on Lobos and may do so on Fuerteventura
Cory's Shearwater	An estimated 3-4,000 pairs on Fuerteventura and probably another 1,000 on Lobos. Very numerous offshore except in mid-winter.
Sooty Shearwater	a few records
Balearic Shearwater	a few autumn records
Manx Shearwater	a few records
Little (Macaronesian) Shearwater	rarely sighted, but may nest both on Fuerteventura and Lobos
White-faced Petrel	a few records from Lobos, where may nest
European Storm Petrel	rarely seen but a few pairs nest on Lobos. Breeding not yet confirmed on Fuerteventura
Madeiran Petrel	May nest Lobos. Has been seen offshore
Leach's Petrel	a few records
Red-billed Tropicbird	a few records, all in recent years
Gannet	common offshore during winter months
Cormorant	a few records
Bittern	a few winter records
Little Bittern	rare migrant
Night Heron	scarce migrant
Squacco Heron	rare migrant
Cattle Egret	previously rare, but has now colonised and probably resident in small numbers.
Little Egret	present all year, can be quite numerous in winter and on passage. A few now nest on Lanzarote, but not proved to do so as yet in Fuerteventura.
Great White Egret	a few records
Grey Heron	present all year round in small numbers
Purple Heron	very scarce passage migrant
Marabou	an African species that has been seen a few times, probably wandering from zoos, but just possibly as a vagrant
White Stork	rare in winter and occasional on passage
Glossy Ibis	a few records
Hadada Ibis	free flying birds in Jandía

Sacred Ibis	a small feral colony in Jandía
Spoonbill	scarce on passage, sometimes winters. Recent colour-ringed birds wintering are from a colony in France
Greater Flamingo	a few records
Greylag Goose	just one record
Ruddy Shelduck	I was lucky enough to find the first for the Canary Islands at the lagoon at Catalina Garcia, Fuerteventura in 1994. It is now well established.
Wigeon	Uncommon winter visitor
American Wigeon	a few records, four wintered in early 1991
Gadwall	a few records
Teal	the commonest winter duck
Green-winged Teal	a single record
Mallard	rare winter visitor
Pintail	rare winter visitor
Garganey	very scarce winter visitor/passage migrant
Blue-winged Teal	a few records
Shoveler	very scarce winter visitor
Marbled Duck	After a long period of absence from the Canary Islands, this duck reappeared in Fuerteventura in the early1990s, when breeding was confirmed and a few were resident throughout that decade. Now a rare visitor.
Pochard	rare winter visitor
Ring-necked Duck	a few records
Ferruginous Duck	a few records
Tufted Duck	very scarce winter visitor
Scaup	a few records
Lesser Scaup	a few records
Honey Buzzard	a single record
Swallow-tailed Kite	a single record
Black Kite	rare passage migrant
Red Kite	a few records
Egyptian Vulture	until recently a common resident, with numbers now recovering slowly due to feeding programme
Short-toed Eagle	a few records
Marsh Harrier	rare winter visitor/passage migrant
Hen Harrier	very scarce winter visitor
Montagu's Harrier	very scarce passage migrant
Dark Chanting Goshawk	one seen 2012
Sparrowhawk	very scarce – status uncertain
Buzzard	now a common resident

Long-legged Buzzard	a few records
Booted Eagle	very scarce passage migrant
Osprey	a scarce resident in the early 80s, but no longer breeds and now a scarce visitor from Lanzarote
Lesser Kestrel	a few records
Kestrel	common resident
Red-footed Falcon	a few records
Merlin	one record
Hobby	rare passage migrant
Eleonora's Falcon	rare visitor, but possibly nested in the past
Lanner	a few reports
Peregrine Falcon	rare winter visitor
Barbary Falcon	scarce resident
Barbary Partridge	rather scarce resident
Quail	very scarce resident
Helmeted Guineafowl	recent report of introduced free-ranging birds
Water Rail	one record
Spotted Crake	scarce passage migrant
Little Crake	rare passage migrant
Baillon's Crake	rare passage migrant
Corncrake	a few records
Moorhen	resident at permanent water
Allen's Gallinule	two records, including one in 2011
Coot	resident at permanent water
Crested Coot	several apparent records
Common Crane	one record of three birds wintering
Crowned Crane	flee-flying birds in Jandía zoo
Houbara Bustard	fairly common resident in suitable habitat
Oystercatcher	rare winter visitor
Canary Islands Black Oystercatcher	extinct
Black-winged Stilt	scarce migrant until 1990s, now a local resident at permanent water
Avocet	rare migrant
Stone Curlew	common but secretive resident
Cream-coloured Courser	fairly common resident in suitable habitat
Collared Pratincole	very scarce spring migrant
Little Ringed Plover	resident at permanent waterbodies and breeds in wet barrancos
Ringed Plover	common winter visitor and passage migrant on coasts
Killdeer	one record
Kentish Plover	fairly common resident on coast and waterbodies

Lesser Sand Plover	one report
Dotterel	rare winter visitor
American Golden Plover	a few records
Pacific Golden Plover	one report
Golden Plover	rare winter visitor
Grey Plover	fairly common winter visitor on coast
Lapwing	rare winter visitor
Knot	very scarce passage migrant
Sanderling	common winter visitor/passage migrant on coast
Semi-palmated Sandpiper	one record
Little Stint	very scarce passage migrant
Temminck's Stint	rare passage migrant
Least Sandpiper	one report
White-rumped Sandpiper	two records
Pectoral Sandpiper	a few records
Curlew Sandpiper	scarce migrant
Dunlin	rather scarce migrant/winter visitor
Ruff	scarce migrant/winter visitor
Buff-breasted Sandpiper	several in autumn 2011
Jack Snipe	a few records
Common Snipe	scarce winter visitor and passage migrant
Woodcock	a few winter records
Black-tailed Godwit	very scarce winter visitor/passage migrant
Bar-tailed Godwit	scarce winter visitor/passage migrant
Whimbrel	common winter visitor/passage migrant on coast
Slender-billed Curlew	one report
Curlew	scarce winter visitor on coast
Spotted Redshank	scarce winter visitor/passage migrant
Redshank	scarce winter visitor/passage migrant
Marsh Sandpiper	a few records
Greenshank	regular winter visitor and passage migrant
Lesser Yellowlegs	one record
Green Sandpiper	regular winter visitor and passage migrant
Wood Sandpiper	scarce migrant
Common Sandpiper	fairly common winter visitor and passage migrant
Spotted Sandpiper	a few records
Turnstone	common winter visitor and migrant on coast
Pomarine Skua	rare passage migrant
Arctic Skua	rare passage migrant
Long-tailed Skua	a few records

Great Skua	rare passage migrant
Mediterranean Gull	a few records
Black-headed Gull	very scarce winter visitor, numbers variable
Slender-billed Gull	has once nested but normally a rare visitor
Audouin's Gull	a few records
Ring-billed Gull	a few records
Common Gull	rare winter visitor
Lesser Black-backed Gull	very scarce, mainly in winter
Yellow-legged Gull	very common resident
Glaucous Gull	a few winter records
Greater Black-backed Gull	rare winter visitor
Kittiwake	rare winter visitor
Gull-billed Tern	irregular and very scarce passage migrant
Caspian Tern	one record
Sandwich Tern	fairly common winter visitor/passage migrant
Roseate Tern	a few records
Common Tern	has nested, but mainly a scarce winter visitor/passage migrant
Arctic Tern	rare migrant
Sooty Tern	a few records
Little Tern	rare migrant
Whiskered Tern	irregular and very scarce passage migrant
Black Tern	irregular and very scarce passage migrant
Razorbill	a few old records
Puffin	rare winter visitor
Black-bellied Sandgrouse	fairly common resident in suitable habitat
Pin-tailed Sandgrouse	a few records
Rock Dove	fairly pure birds are common on coastal cliffs etc but birds inland are more often Feral Pigeons
Woodpigeon	rare winter visitor
Collared Dove	now an abundant resident, but entirely absent until 1990s.
African Collared Dove	introduced and now widely established as feral population in small numbers
Turtle Dove	fairly common but local summer visitor and passage migrant. Some perhaps winter in south.
Laughing Dove	scarce and local resident recently colonised
Monk Parakeet	a small feral breeding population in Jandía

Ring-necked Parakeet	free-flying birds in Jandía, but not established
Great Spotted Cuckoo	rare winter visitor/passage migrant
Cuckoo	scarce migrant
Barn Owl	very scarce resident
Scops Owl	rare migrant
Eagle Owl	one record, probably an escaped bird
Tawny Owl	one record
Long-eared Owl	rare resident
Short-eared Owl	rare winter visitor
Nightjar	a few records
Red-necked Nightjar	one record
Plain Swift	scarce and local resident
Common Swift	fairly common passage migrant
Pallid Swift	common resident
Alpine Swift	rare migrant
Little Swift	a few records
White-rumped Swift	one report
Kingfisher	a few records
Bee-eater	scarce migrant
Roller	rare migrant
Hoopoe	common resident
Wryneck	scarce migrant
Hoopoe Lark	a few records
Dupont's Lark	one record of possible breeding pair
Calandra Lark	a few records
Short-toed Lark	rare migrant but possibly overlooked
Lesser Short-toed Lark	very common resident
Skylark	scarce winter visitor
Plain Martin	a few reports
Sand Martin	scarce migrant
Crag Martin	a few records
Swallow	common passage migrant and very scarce winter visitor
Red-rumped Swallow	very scarce migrant
House Martin	common passage migrant
Richard's Pipit	one record
Tawny Pipit	scarce passage migrant and rare winter visitor
Berthelot's Pipit	abundant resident
Tree Pipit	common passage migrant
Meadow Pipit	scarce winter visitor
Red-throated Pipit	rare migrant
Olive-backed pipit	first records in November 2012
Water Pipit	one recent record
Rock Pipit	a few records
Yellow Wagtail	fairly common passage migrant
Grey Wagtail	scarce winter visitor

White Wagtail	scarce winter visitor
Rufous Bush-chat	a few records
Robin	scarce winter visitor
Thrush Nightingale	a few reports
Nightingale	scarce migrant
Bluethroat	rare migrant
Black Redstart	scarce winter visitor
Redstart	scarce migrant
Whinchat	fairly common migrant
Fuerteventura Chat	common but local resident
Stonechat	rare winter visitor
Wheatear	fairly common migrant
Black-eared Wheatear	very scarce migrant
Desert Wheatear	a few winter records
Blue Rock Thrush	a few records
Ring Ouzel	a few winter records
Blackbird	rare, but has apparently nested in the past
Fieldfare	rare winter visitor
Song Thrush	scarce winter visitor
Redwing	very scarce winter visitor
Grasshopper Warbler	a few records, probably overlooked
Savi's Warbler	one report
Aquatic Warbler	a few records
Sedge Warbler	very scarce migrant
Reed Warbler	very scarce migrant
Great Reed Warbler	a few records
Eastern Olivaceous Warbler	one report
Western Olivaceous Warbler	rare migrant
Icterine Warbler	one report
Melodious Warbler	scarce migrant, mainly in autumn
Tristram's Warbler	two records
Spectacled Warbler	common resident
Subalpine Warbler	scarce migrant
Menetries' Warbler	one report
Sardinian Warbler	common but local resident
African Desert Warbler	a few records
Orphean Warbler	a few records
Lesser Whitethroat	a few reports
Whitethroat	very scarce migrant
Garden Warbler	scarce migrant
Blackcap	fairly common winter visitor and passage migrant
Yellow-browed Warbler	rare winter visitor and autumn migrant
Hume's Yellow-browed Warbler	one report
Bonelli's Warbler	scarce migrant
Wood Warbler	fairly common migrant
Chiffchaff	common winter visitor

Iberian Chiffchaff	rare winter visitor
Willow Warbler	common migrant
Spotted Flycatcher	scarce migrant, mainly in autumn
Red-breasted Flycatcher	a few records
Collared Flycatcher	one record
Pied Flycatcher	fairly common migrant
African Blue Tit	fairly common but local resident
Great Tit	one record
Golden Oriole	very scarce migrant
Isabelline Shrike	one report
Southern Grey Shrike	common resident
Woodchat Shrike	scarce migrant
Brown-necked Raven	several reports, but all probably refer to the following species
Raven	common resident
Red-vented Bulbul	introduced and probably not fully established but breeds wild in at least some of the resorts
Starling	very scarce winter visitor
Spotless Starling	one report
Spanish Sparrow	common resident
Rock Sparrow	a few records
Hawfinch	one record
Chaffinch	a few records
Brambling	a few records
Serin	a few records
Canary	introduced resident now common in Betancuria mountains, sometimes seen in other areas
Greenfinch	a small resident population in the late 70s and early 80s but apparently now extinct
Goldfinch	scarce resident
Siskin	rare winter visitor
Linnet	common resident
Trumpeter Finch	common resident
Blackpoll Warbler	one record
Ortolan Bunting	scarce passage migrant
Cretzschmar's Bunting	adult male seen by Howard Taffs, Dave Shirt and myself in 1980.
Little Bunting	a few records
Reed Bunting	one record
Corn Bunting	local resident

Key References

Bannerman, D. A. 1922 *The Canary Islands: their history, natural history and scenery.* Gurney and Jackson, Edinburgh

Clarke, T. 2006 *Birds of the Atlantic Islands.* Christopher Helm, London.

Clarke, T. and Collins, D. 1996 *A Birdwatchers' Guide to the Canary Islands.* Prion Ltd, Huntingdon.

Lorenzo, J. A. (ed.) 2007 Atlas de las aves nidificantes en el archipiélago canario (1997-2003). Direccíon General de Conservación de la Naturaleza-Sociedad Española de Ornitología. Madrid

Martín, A. et al. 1989 *Las Aves Marinas de Canarias.* La Garcilla 73:8-11

Moreno, J. M. 1988 *Guía de las Aves de las Islas Canarias.* Editorial Interinsular Canaria, Santa Cruz de Tenerife. ISBN: 84-86733-05-7

Rodriguez Delgado, O. 2005 *Patrimonio Natural de la Isla de Fuerteventura.* Gobierno de Canarias. ISBN: 84-7926-391-1

Appendix Two

Plants of Fuerteventura

The following is not a complete list of the plants found in Fuerteventura, but includes all the commoner species that are likely to be found on a trip to the island, plus all the endemic species.

The scholarly text by Stephan Scholz in the book *Patrimonio Natural de las isla de Fuerteventura*, published in 2005, lists 678 as having been confirmed in Fuerteventura. Of these, 13 are endemic to Fuerteventura, 33 are endemic to the eastern Canary Islands, and another 42 are endemic to the Canary Islands. I have also referred to the earlier work by Gunter Kunkel *Las Plantas vasculares de Fuerteventura (Islas Canarias), con especial interés de las forrajeras* which was published in 1977 and remains an important source of information on the distribution and status of plants on the island.

For introduced species I have also referred to the paper by Dietmar Brandes & Ketrin Fritzsch 'Alien plants of Fuerteventura, Canary Islands', published in 2002. According to them, 780 species have been recorded on the island, of which at least 119 are alien and some 150 further species are of Mediterranean and/or North African origin and probably also introduced. This paper, which can be viewed in full on the internet, has checklists identifying the origin of alien species and photos of around thirty alien species.

The Bramwells' guide to the wild flowers of the Canary Islands is very useful for identifying the endemic plants, and many others, though by no means all, are included in flower guides to the

Mediterranean. The website www.floradecanarias.com has excellent photographs of many of the species with a few notes (in Spanish) giving basic identification features.

★★★   denotes species (or subspecies) endemic to Fuerteventura

★★   denotes species (or subspecies) that are endemic to Fuerteventura and Lanzarote

★   denotes species that are endemic to the Canary Islands

I have provided English common names where they are available, and a few of the more important local Spanish names for key species that have no English name.

Gymnosperms

Pinus canariensis	**Canary Islands Pine** Small numbers in the Betancuria forest
Pinus halepensis	**Aleppo Pine** The most commonly planted species in the Betancuria forest
Pinus radiata	**Monterey Pine** Also planted in the Betancuria forest

DICOTYLEDONS

AIZOACEA

Aizoon canariense	Very common. Prostrate creeping annual, small yellowish fls.
Mesembryanthemum crystallinum	**'Barrilla'** Very common. Flat, crystal covered leaves and white star fls (Plate 15).
Mesembryanthemum nodiflorum	**'Cosco'** Abundant annual that often forms reddish patches.

AMARANTHACEA

Bosea yervamora

Rare in Jandía mountains. A leathery-leaved shrub.

APIACAE (UMBELLIFERAE)

Astydamia latifolia

Big fleshy leaves. Coastal rocks, especially north of Cotillo.

**Bupleurum handiense*

Shrubby umbelifer with yellow flowers, only on inaccessible cliffs of Jandía.

**Ferula lancerottensis*

Previously thought to occur only in Lanzarote, but also in Jandía mountains.

**Rutheopsis herbanica*

Yellow flowered umbelifer in the high mountain areas and malpais at La Oliva.

ASCLEPIADACEAE

Calotropis procera

'Ginny Plant'
Established around Giniginámar.

Carraluma burchardii

Succulent. The subsp in Fuerteventura is endemic to the Eastern Canary Islands (Plate 23)

ASTERACEAE (COMPOSITAE)

Andryala glandulosa

'Ropa Vieja' or Old Man's Clothes
Woolly plant of mountains/cliffs.

***Argyranthemum winteri*

Rare daisy-flowered shrub found only on the Jandía cliffs.

Asteriscus schultzii

Small shrub with large pale yellow flowers found only in the northern part of the island. Previously considered as endemic to the Eastern Islands but also know from Morocco.

***Asteriscus sericeus* '

Characteristic shrub of mountain zones, with big yellow flowers (Plate 16)

Calendula arvensis (includes *C. aegyptiaca*)

Abundant annual throughout the island.

***Carduus bourgeaui*

Thistle on northern slopes of Jandía mountains only, where locally abundant.

Carduus clavulatus

Jandía cliffs and perhaps a few other mountain cliffs.

Chrysanthemum coronarium

Crown Daisy
Common on roadsides, cultivations etc.

**Crepis canariensis (Lactuca herbanica)*	**Canary Islands Hawksbeard**. Common on cliffs and amonst rocks in the mountains.
Dittricha (Inula) viscosa	Fairly widespread ruderal species on roadsides, in barrancos etc.
Filago (Logfia) gallica	**Narrow-leaved Cudweed** In sandy places
Filago pyramidata	**Broad-leaved Cudweed** Cultivated areas in the north
Ifloga spicata	The common Cudweed on the island, found in a wide range of habitats.
★Kleinia neriifolia (Senecio kleinia)	**'Verode'** Common succulent shrub throughout the island.
Launaea arborescens	**'Aulaga'** Bush with soft spines, abundant throughout the island.
Launaea nudicaulis	The common yellow composite in sandy places.
***Onopordon nogalesii*	Purple flowered thistle: 100 individuals in Barranco de Vinámar (below Pico de la Zarza).
Pallenis hierochuntica (Asteriscus pygmaeus(aquaticus)	**Rose of Jericho** Very common
Phagnalon purpurascens	Widespread on rocks in mountains and barrancos.
Phagnalon rupestre	In similar places to the preceding species
Pulicaria burchardii	Previously considered endemic to Fuerteventura (a few places in Jandía). Now known also from western parts of the Sahara and the Cape Verde Islands.
**Pulicaria canariensis*	A showy, yellow flowered species. On rocks and cliffs (especially sea cliffs) in a few scattered locations. (Plate 11)
**Reichardia famarae*	Only on the Jandía cliffs.
Reichardia tingitana	Widespread on verges etc mainly in the north. Dark purple centres to flowers.
**Senecio bollei*	Restricted to the Jandía cliffs, where it is common.
Senecio glaucus (coronopifolius)	Widespread species with ray florets. Very variable leaves.
Silybum marianum	**Milk Thistle** Northern and central mountains, and abandoned cultivations.

*Sonchus leptocephalus
   (Atalanthus pinnatus)
          A lanky, shrubby species on Montaña Cardon.

Sonchus pinnatifidus
          Rare plant on mountain cliffs in the south of the island. Short, woody stem.

Sonchus tenerrimus
          A widespread Mediterranean species in cultivated areas. Leaves deeply pinnate.

Urospermum picroides
          Widespread ruderal species

**Volutaria bollei
          One site in the the Rio Palmsa valley.

BORAGINACEAE

*Ceballosia (Messerschmidia) fruticosa
          A shrub, rare on this island, recently seen in the Cardon area.

*Echium bonnetii
          Widespread and abundant. The subspecies fuerteventurae is endemic.

*Echium decaisnei
          A handsome white flowered shrub. Subspecies purpuriense, formerly known as Echium famarae is found on mountain cliffs, mostly in Jandía. The nominate form from Gran Canaria is introduced and occurs here and there on roadsides.

***Echium handiense
          A rare, blue-flowered sub-shrub found only on the high cliffs of Jandía.

Heliotropium ramosissimum
          Widespread and very common perennial, especially in sandy areas.

BRASSICACEAE (CRUCIFERAE)

Cakile maritima
          Sea Rocket
          Common in coastal sands

Carrichtera annua
          Common in abandoned cultivations etc.

***Crambe sventenii
          Shrubby species on mountain cliffs at the south of the island (not Jandía).

*Erucastrum canariense
          Common yellow flowered species in rocky places, especially mountain areas.

Hirschfeldia incana
          Frequent on roadsides etc in the northern half of the island.

Lobularia canariensis (intermedia)
          Sub-shrub of rocky places, especially in hilly areas.

Lobularia lybica
          Small annual – widespread and common, especially in sandy areas.

Mattiola fruticulosa (bolleana)
          A showy plant with large mauve and white flowers, common on rocky ground near the south coast, but also occurring here and there in other places.

M. bolleana	Was long considered an endemic of the Eastern islands, but is now included in this widespread Mediterranean species.
Mattiola parviflora	More ruderal species, with much smaller flowers and densely felted leaves.
Moricandia arvensis	Frequent and conspicuous ruderal in the sw of the main island. Mauve fls.
Notoceras bicorne	An inconspicuous plant but abundant and widespread.
Sisymbrium erysimoides	Ruderal with spreading seed pods: widespread along roads, abandoned cultivations etc.
Sisymbrium irio	**London Rocket** As per the previous species, but with pods parallel to stem.

CACTACEAE

Opuntia dillenii	Introduced but now naturalised over large parts of the island.
Opuntia ficus-indica	**Prickly-pear Cactus** Extensive plantations, and naturalised in places.

CAMPANULACEAE

Campanula occidentalis (dichotoma)	An uncommon, large flowered species found in mountains.

CARYOPHYLLACEAE

Gymnocarpos decandrus (salsoloides)	Succulent leaved species found in coastal areas of Jandía.
***Herniaria hartungii*	A poorly known species from Jandía.
Minuartia geniculata	Widespread Mediterranean species found in sandy areas.
***Minuartia platyphylla*	The upper part of the Jandía cliffs. Leaves broad.
Minuartia geniculata	Widespread Mediterranean species found in sandy areas.
***Minuartia webii*	Scarce but widely distributed on rocky ground.
Polycarpaea divaricata	A plant of the high cliffs in Jandía.
Polycarpaea nivea	A fleshy leaved, silvery hairy shrublet abundant in sandy areas such as at Corralejo, Cotillo, Lajares and Jandía isthmus. Also on some plains.

Polycarpon tetraphyllum	**Four-leaved Allseed** Widespread ruderal species.
Spergula arvensis	**Corn Spurrey** Widespread in cultivations etc.
Spergularia diandra	**Lesser Sand-spurrey** Widespread. Petals equal sepals.
Spergularia media	**Greater Sea-spurrey** The common sea-spurrey on the island.

CELASTRACEAE

★Maytenus canariensis	Stunted evergreen trees grow on the Jandía cliffs. Pale berries.

CHENOPODIACEA

Arthrocnemum macrostachyum (fruticosum)	The dominant shrub in coastal saltmarshes. Branched fleshy stems give something of the appearance of ostrich feathers.
Atriplex glauca	The common small atriplex shrub in sandy areas
Atriplex halimus	Common, larger shrub. Pale stems and dense 'knots' of leaves – coastal sands.
Beta maritima	**Sea Beat** Widespread ruderal species
Chenoleoides	*(Chenolea) tomentosa* Common shrub with densely felted leaves, mainly on coastal plains.
Chenopodium album	**Fat Hen** Common and widespread ruderal species.
Chenopodium murale	**Nettle-leaved Goosefoot** Common and widespread ruderal. Toothed leaves.
Patellifolia (Beta) patellaris	Widespread and common, especially on the coast. Leathery leaves.
★Patellifolia (Beta) webbiana	Rare on coastal rocks. Dagger-shaped or linear leaves.
★Salsola divaricata (longifolia)	Shrub of sandy places. Hairless cylindrical leaves, pale winged fruits.
Salsola vermiculata	Abundant shrub everywhere. Fruits with conspicuous pink/red wings.
Suaeda mollis	Widespread shrub in the coastal zone.
Suaeda vera	**Shrubby Seablight** Dominant in permanently wet barrancos, also in saltmarshes.

Traganum moquinii	Large shrub forming large patches in coastal sand dunes, especially Corralejo.

CISTACEAE

Helianthemum canariense	Common yellow flowered subshrub with densely hairy leaves.
**Helianthemum thymiphyllum*	Like *H. canariense* but hairless, bright green leaves. Mountains of Betancuria and Jandía.

CONVOLVULACEAE

Convolvulus althaeoides	**Mallow-leaved Bindweed** Conspicuous and common in the Betancuria hills, but also ruderal in other areas.
Convolvulus caput-medusae	**Medusa's-head Bindweed** Rare cushion forming shrublet of coastal rocks (south-west) and the sandy Jandía isthmus.
Convolvulus floridus	A tall shrub. Rare in Jandía.
Convolvulus siculus	**Small Blue Convolvulus** Widespread in open rocky ground. Blue flowers.
Cuscuta approximata ssp. *episonchus*	**Dodder** Common, forming string-like patches in Launaea bushes.

CRASSULACEA

**Aeonium balsamiferum*	Succulent shrub native to Lanzarote and probably introduced in Fuerteventura, where it is found amongst Opuntia and Agave plantations in the north and centre.
***Aichryson bethencourtianum*	Rare – Jandía cliffs. Short-stemmed leaves. Flowers have 8 petals.
Aichryson laxum	Jandía cliffs and Cardon. Flowers have 9-12 petals.
Aichryson pachycaulon	Rare plant of the Jandía cliffs. Small flowers with 6-8 petals.
**Aichryson tortuosum*	Cliffs in the northern part of the island. Leaves sessile.
Monanthes laxiflora (microbotrys)	Cliffs and rocky slopes in the mountains. Scarce

Umbilicus gaditanus (horizontalis)

Navelwort
Widespread and common on rock
ledges etc.

CUCURBITACEAE

Citrullus colocynthus

Bitter Apple
Widespread and common gourd
forming prostrate trailing plant in
barrancos and other places where water
stands after rains.

EUPHORBIACEAE

Euphorbia balsamifera

'Tabaiba Dulce' Succulent shrub 2m
tall. Larger and darker looking than
E. regis-jubae. Forms extensive stands in
a few places, especially on south-facing
mountain slopes.

*Euphorbia canariensis

'Cardón' (Plate 20)
Forming dense clumps of tall 'organ-
pipes' in a few places. Below Montaña
Cardón and in Jandía

***Euphorbia handiensis

A few colonies of this cactus-like plant
on south facing slopes of western Jandía.

Euphorbia paralias

Sea Spurge
Coastal sands. Especially common
south-east of Corralejo.

Euphorbia regis-jubae (obtusifolia)

'Tabaiba Amarga' (Plate 17)
This succulent shrub is a widespread
and conspicuous element of the island's
flora. Smaller than E. balsamifera with
conspicuous yellowish floral bracts.

Mercurialis annua

Annual Mercury
Widespread ruderal species in malpais,
rocky hill slopes etc.

Ricinus communis

Castor-oil Plant Widespread
introduced species in damper places
such as barrancos.

FABACEA (LEGUMINOSAE)

*Asphaltium bituminosum
(Psoralea bituminosa)*

Pitch Trefoil
The typical form is a ruderal plant on
roadsides etc in the centre and north,

but the form *albomarginata*, which
can have woody stems up to 4cm across,
grows on the cliffs of Jandía
and Montaña Cardón.

Astragalus hamosus
Milk-vetch
Common on plains, especially in north.
Yellow flowers and very curved pods.

**Astragalus mareoticas* var *handiensis*
Has pinkish flowers and less curved
pods. Less common.

Hippocrepis multisiliquosa
Widespread Horseshoe vetch with
yellow flowers and very characteristic
notched pods curled into a circle.

Lotus glinoides (arabicus)
Small plant not uncommon near coast.
Pinkish flowers.

Lotus lancerottensis
Common and widespread. Yellow
flowers (Plate 12).

Medicago laciniata
Widespread. Single or paired flowers,
deeply toothed or lobed leaves.

Medicago littoralis
Widespread in coastal areas.

Medicago minima
Bur Medick
Fairly widespread. Densely hairy.

Medicago polymorpha (nigra)
Toothed Medick
Widespread and common. Deeply
toothed stipules.

Medicago sativa
Lucerne
Widely cultivated and sometimes found
growing wild.

Melilotus sulcatus
Furrowed Melilot
Widespread ruderal species.

***Ononis christii*
Only on the high cliffs of Jandía. Small
shrub with large, pale pink flowers.

**Ononis catalinae (hebecarpa)*
Fairly widespread, especially Jandía and
Lobos. Yellow flowered annual, recently
described as a species distinct from *O.
hebecarpa*.

Ononis hesperia (natrix)
Large Yellow Restharrow
Common in sandy places. Sticky yellow
flowered sub-shrub.

Ononis serrata
Widespread in coastal areas and north
ern malpais. Showy pink-flowers.

Scorpiurus muricatus
Scorpion Vetch
Widespread in cultivations etc.

Vicia benghalensis
Cultivated areas, roadsides etc. in centre
and Jandía. Purple flowers.

Vicia sativa
Common Vetch
Cultivations etc in north and centre.

FRANKENIACEAE

Frankenia capitata (laevis)

Sea Heath
Widespread and common especially on coast.

Fumaricacea
Fumaria muralis

Common Ramping Fumitory
common ruderal

GERANIACEAE

Erodium chium

Frequent ruderal species in centre and north. Bluntly lobed oval leaves.

Erodium cicutarium

Common Storksbill
Frequent ruderal species. Bi-pinnate leaves.

Erodium malacoides

Mallow-leaved Storksbill
Widespread and common ruderal species. Leaves oblong, barely lobed.

LAMIACEAE (LABIATAE)

Ajuga iva

Very common and widespread. Sprawling plant with yellow flowers.

Lavandula canariensis (multifida)

Shrub with bright green leaves on mountain cliffs. Fls. blue

Lavandula pinnata

Greyish sub-shrub with purple fls. Rare in southern Fuerteventura.

Marubium vulgare

White Horehound
Widespread and abundant ruderal, especially in mountains.

Micromeria varia

Common on mountain rocks.

Salvia canariensis

Only around La Oliva. Probably a garden escape. Triangular leaves, purple flrs.

***Salvia herbanica*

Mountains in the SE only. Dwarf shrub with linear leaves.

Salvia vernenaca

Wild Clary
Common and widespread ruderal in the north. Lobed leaves

**Sideritis (Leucophae) pumila (massoniana)*

Jandía cliffs. Shrublet with velvety, heart-shaped leaves. Pale yellow fls.

MALVACEAE

*Lavatera aceriflolia
A few individuals on one cliff only.
Shrub with sycamore-like leaves.

Lavatera cretica
Smaller Tree Mallow
Ruderal species in north and centre.

Malva parviflora
Least Mallow
Common and widespread ruderal.

MIMOSACEAE

Acacia cylcops
Shrub planted and naturalised in the
Betancuria forest. Wavy pods.

Acacia farnesiana
Needle Bush
Shrub with natural regeneration in
places. Thorny branches.

MORACEAE

Ficus carica
Fig Tree
Cultivations, not reproducing naturally.

OLEACEAE

*Olea cerasiformis
Canary Islands Olive
A few stunted individuals survive on
mountains. Scattered bushes remain in
the Betancuria area.

OROBANCHACEAE

Csitanche phelipaea
Common in coastal sands. Stout, yellow
flowered broomrape.

Orobanche ramosa
Branched Broomrape
Widespread. Often branched.
Lax-flowered spikes.

OXALIDACEAE

Oxalis corniculata
Yellow Sorrel
Widespread in cultivations, roadsides,
gardens etc.

Oxalis pes-caprae
Bermuda Buttercup
Cultivations, gardens etc in Betancuria
area.

PAPAVERACEAE

Papaver hybridum

Rough Poppy
Widespread. Generally the common poppy on the island.

PLANTAGINACEAE

Plantago afra

Widespread and common on plains etc. Branched. Leaves linear, opposite.

Plantago amplexicaulis

Widespread ruderal species. Woody stock.

Plantago ciliata

Widespread and common. Densely hairy leaves and clusters of woolly fls.

Plantago coronopus

Buckshorn Plantain
Common in coastal sands and plains. Pinnate leaves.

Plantago lagopus

Ruderal species in cultivations etc in north. Like Ribwort Plantain.

Plantago ovata

Widespread ruderal species. Rosette of long, lanceolate leaves.

PLUMBAGINACEAE

**Limonium borgaeui*

Very rare on Jandía cliffs. Winged flower stems and pubescent leaves.

Limonium ovalifolium

Coastal dunes on Lobos. Small plants with rosette of basal leaves.

**Limonium papillatum papillatum*

Rocky/gritty coastal sites on the north coast and Jandía. Has masses of pink flowers on zig-zaging inflorescences (Plate 9).

**Limonium puberulum*

A smaller version of *L. borgaeui*. In Fuerteventura, known only from the mouth of Barranco de la Torre, but perhaps now extinct on the island.

Limonium tuberculatum

Small shrub with pink fls. Common on Lobos. Rare in Fuerteventura.

POLYGONACEAE

Rumex bucephalophorus

A common and widespread species. Seeds spiny winged on drooping stalks.

**Rumex lunaria*

In and around houses where planted. Shrub with ovate leaves.

Rumex vesicarius rhodophysa

Widespread and common. Large, red tinged, inflated fruits.

PRIMULACEAE

Anagallis arvensis

Scarlet Pimpernel
Abundant in many habitats. Flowers usually blue.

RANUNCULACEAE

Adonis microcarpa

Small Pheasant's-eye
Widespread but local in mountains.

Ranunculus cortusifolius

Widespread in rocky places in mountains and in malpais at La Oliva. Robust perennial. The only buttercup on the island.

RESEDACEAE

****Reseda lancerotae (crystallina)*

Common in sandy areas but also rocks in mountains. Yellow flowers.

RHAMNACEAE

*Rhamnus crenulata

Jandía cliffs. Evergreen shrub.

ROSACEAE

Rubus bollei

'Zarza'
Inaccessible ledges on Jandía cliffs.

RUBIACEAE

*Plocama pendula

Very rare. A shrub with thread-like hanging leaves.

Rubia fruticosa

Madder
Widespread but local. Malpais, mountain rocks etc. Scrambling shrub with whorles of very prickly leaves. Berries usually pale.

SCROPHULARIACEAE

*Campylanthus salsoloides

Mountain cliffs throughout the island. Small shrub with fleshy, linear leaves and pinkish/bluish flowers.

Kickxia sagittata (heterophylla)	Widespread and common in all areas. A yellow flowered toadflax (Plate 13).
Scrophularia arguta	Widespread in rocky areas. The only figwort on the island.

SOLANACEAE

Datura innoxia	Widespread in barrancos and damp places.
Datura stramonium	**Thornapple** Less common than the preceding species.
Hyoscyamus albus	**White Henbane** Scarce in nitrogen-rich places.
Lycium intricatum	**Espino** A widespread and abundant trailing shrub. Characteristic spiny plant with tubular mauve flowers and red berries.
Nicotiana glauca	**Shrub Tobacco** Abundant and characteristic plant, especially along roadsides and in barrancos. Tall woody species with drooping yellowish tubular flowers.
Solanum nigrum	**Black Nightshade** Common and widespread ruderal species.

TAMARICACEAE

Tamarix canariensis	**Canary Islands Tamarisk ('Tarajal')** This tall shrub forms the dominant vegetation in many of the larger barrancos.

URTICACEAE

*Forsskaolea (Forsskahlea) angustifolia	Widespread and common in disturbed ground etc. Robust perennial with prickly leaves.

ZYGOPHYLLACEAE

Fagonia cretica	Widespread and common. Sprawling with atractive, five petalled, bright magenta fls.

Zygophyllum fontanesii

Sea Grape
Common fleshy-leaved shrub on coastal
rocks and sand.

Monocotyledons

AGAVACEAE

Agave americana

Century Plant
Planted on boundaries and naturalising.
Stout leaves.

Agave fourcroydes

Cultivated in plantations for sisal
production and regenerating naturally.
Leaves strap-like.

Agave sisalana

As per the preceding species. Flowers
produce bulbils.

AMARYLLIDACEAE

★Pancratium canariensis

Sea Daffodil
Rare on high cliffs.

ARACEAE

Aricarum vulgare

Friar's Cowl
Fairly common on rock ledges in high
mountains, where the rather large, heart
shaped leaves are commonly found, but
the extraordinary flowers very much
less so.

Arum itallicum

Large Cuckoo-pint
Uncommon in the northern of the
island.

ARECACEAE

★Phoenix canariensis

Canary Islands Palm
Widely cultivated and grows wild in
places.

Phoenix dactylifera

Date Palm
Cultivated and naturalised in the south
of the island.

CYPERACEAE

Cyperus capitatus (kalli)
Abundant in coastal sands and other sandy places.

IRIDACEAE

Romulea columnae
Sand Crocus
Widespread in rocky places in mountains.

JUNCACEAE

Juncus acutus
Sharp Rush
Found in a few barrancos, but very scarce.

Juncus bufonius
Toad Rush
Wet places, eg. the stream below Vega de Rio Palmas.

LILIACEAE

Alium vineale
Crow Garlic
Common in rocky places, especially in mountains throughout the island. Fls often bulbils only, overtopped by long papery bract.

Aloe vera
Recently planted as a crop and now used widely in landscaping.

**Androcymbium psammophilum psammophilum
Common in consolidated sand in the north (Corralejo-Cotillo-Lajares). Stemless white fls at centre of star-like basal rosette.

*Asparagus arborescens
Very rare on cliffs, mainly Jandía. A tall, erect shrub.

**Asparagus nesiotes purpuriensis
Fairly widespread in rocky places. Scrambling shrub with short, blunt spines and smooth stems.

Asparagus pastorianus
Common in malpais in north and rocky montane places in centre. Scrambling shrub with sharp spines.

*Asparagus umbellatus
Cliffs in Jandía. Like A. nesiotes but stems papillose.

Asphodelus fistulosus (tenuifloius)
Hollow-leaved Asphodel
Very common and widespread in lowlands.

Asphodelus ramosus (microcarpus) | **Common Asphodel**
Common and widespread in mountains. Much larger and more robust than the preceding.

Dipcadi serotinum | Widespread but not common. Like a brown-flowered bluebell.

Drimia (Urginea) maritima | **Sea Squill**
Widespread and common, forming large patches on hillsides etc. Large leaves that die back before tall spike of white fls appears.

POACEAE (GRAMINEAE)

Arundo donax | **Giant Reed**
In barrancos, near habitations etc. Bamboo-like stands.

Avena barbata | **Bearded Wild Oat**
Common on plains etc in north and centre.

Avena canariensis | Common and widespread. Lax, one sided, nodding inflorescence.

Cynodon dactylon | **Bermuda grass**
Widespread in various habitats. Spikes 3-6 in a hand-like fan.

Lamarckia aurea | **Golden Dog's-tail**
Widespread and common. Dense, one sided spikes that turn golden as they mature.

Pennisetum setaceum | An introduced ornamental grass naturalising in places. Dense cylindrical spikes more than 10cm long feathery appearance due to 3cm long silvery hairs on fls.

Phalaris caerulescens | Widespread in cultivated ground etc. Very like following species in general appearance but fls usually purplish.

Phalaris canariensis | **Canary Grass**
Widespread in cultivated areas etc.

Trisetaria lapalmae | A recently described species apparently fairly widespread on the island.

APPENDIX THREE

Mountains of Fuerteventura

A list of mountains and subsidiary peaks in the island, including all above 500m and other significant ones.

North-east

Montaña de Escanfraga	529m
Montaña de Enmedio	531m
Morro Tabaiba	527m
Morro de la Majada	503m
Morro de los Rincones	503m
Morro Carnero	510m
La Muda	689m
Morros Altos	565m
Pico de Don David	554m
Pico del Sabio	563m
Morro de la Pila	468m
Aceitunal	688m
Montaña de la Caldera	514m
El Piquito	545m
Morro de Agua Salada	509m
Morro de Aguas Blandas`	524m
Morro de Facay	463m
Cuchillos	627m

Pico de la Fortaleza	595m
Morro Bermejo	616m
Morro de la Galera	537m
Solanas de Casillas	527m
Morro de la Atalaya	563m
Morro de las Piteras	512m
Castillejo Grande	495m
Cerro de Temejereque	511m

SIGNIFICANT NORTH-EAST HILLS

Montaña Arena	420m
Montaña Tindaya	401m
Montaña de Pedra Sal	497m
Montaña de San Andres	458m
Montaña de Tesjuates	439m

South-east

Rosa del Taro	596m
Morro de la Cochina	543m
Morro de los Asientos	485m
Morro de las Tinajas	545m
Montaña del Campo	542m

SIGNIFICANT SOUTH-EAST HILLS

Morro Pinacho	483m
Morro de Valle Corto	459m
Buenavista	417m
Montaña Gairia	463m
Agudo	497m
Morro de las Casas	408m
Morro de los Halcones	433m
Atalaya de Pozo Negro	439m

Caldera de Jacomar	435m
Vigan	464m

Betancuria Massif

La Atalaya	538m
La Torecilla Chica	535m
Morro de la Fuente Vieja	666m
Morro de Graman	656m
Morro del Corralete	508m
Morrete de Cerdena	599m
Morro de la Cruz	674m
Morro de Velosa	675m
Montaña Tegu	645m
Morro Janana	674m
Morro de la Vieja	633m
Morro de Tabagoste	624m
Montaña Atalaya	724m
Morro Tabaiba	670m
Gran Montaña	708m
Morro Jorjado	679m
Pico Alto	609m
Fenduca	609m
Carbon	608m

South-west

Cardon	694m
Espigon de Ojo Cabra	661m
Degollado de la Galera	602m
Sisacumbre	528m
Melindraga	631m
Degollado del Risco	489m
Montaña de la Fuente	497m
Montaña Hendida	484m

SIGNIFICANT SOUTH-WEST HILLS

Caracol	467m

Jandía

Morro de la Burra	518m
Pico de la Zarza	812m
Pico de Mocan	792m
Morro del Joaro	621m
Morro del Cavadero	734m
Fraille	686m
Morro de la Habana	528m
Pico de la Camella	608m

SIGNIFICANT JANDÍA HILLS

Montaña Aguda	435m

General index

Index of Birds

Index of higher plants